RAINFOREST of tropical Australia

Text and Photography
by Damon Ramsey, B.Sc.

Ecosystem Guides

National Library of Australia Cataloguing-in-publication data.
Ramsey, Damon.
Rainforest of tropical Australia
Cairns, Qld: Ecosystem Guides, 2008
ISBN 9780975747032
Includes Index. Bibliography
Plant ecology - tropics, Rainforest ecology - Australia
577.340994

"Rainforest of Tropical Australia"

ISBN number 976-0-9757470-3-2.
Published by Ecosystem Guides
Cairns, Queensland, Australia.
Various desktop editions printed from 2002,
This edition printed 2007.
Typeface Arial Narrow with Adobe InDesign.
Printed by Yes Printing, Sydney, Australia

The author cannot be responsible for readers getting lost in the jungle, scratched by wait-a-while, getting their lunch stolen by brush turkeys, getting sick from gobbling bush tucker, or any other mishaps derived from following the information in this manual. Be responsible for your own life!

For any enquiries about services, please contact the email address below:
damon@educational-tours.com.au

Cover photographs
Main background photo: Tropical rainforest around Mt. Hemmant, Daintree.
Top left: White-lipped tree frog *Litoria infrafrenata*, Mission Beach.
Top right: Southern Cassowary, *Casuarius casuarius*, Mission Beach.
Bottom left: Flowers of *Alphitonia*, Atherton Tablelands.
Bottom right: Lacewing butterfly, New Guinea.

Educational Tours Australia

Educational Tours Australia is a company devoted to interpreting Australia and its natural and cultural history. We do this through: a free educational information-based website, a series of ecosystem guides, hiring out biologist guide Damon Ramsey for tours & lectures, wildlife & nature tours and study tours & programmes.

For enquiries: ***www.educational-tours.com.au***

Nature & Wildlife tours

ETA conducts tours that explore Australia's unique plants and animals, personally guided by biologist and Ecosystem Guides author Damon Ramsey. We explore the great Barrier Reef, rainforest, outback and mountains and search for kangaroos, kookaburras, crocodiles, flowers, butterflies, tropical fish and hundreds of other species.

Study tours and programs

ETA arranges trips for a range of interests and ages, from primary kids who want a fun trip to more serious college level biology studies. The aim of our biology study tours is to allow students the opportunity to study biology in an outdoor setting. This study tour is made up of short lectures, field-based nature walks, surveys and wildlife spotting and identification. The programme can also allow lectures by your own specialists during the trip. A detailed itinerary can be found on our website. The programme has the potential to fit into various credited courses and or to stand alone as a study experience. It can also slot in as part of a larger itinerary within Australia or the Pacific, of which we can assist with ideas and bookings. We can also arrange variations of these study tours. We can organize different workshops from plant identification to frog surveys and present a range of lecture presentations from geology, plants, mammals or Aboriginal art.

Contents

Introduction

Aims of the Ecosystem Guide series

This series is an attempt at 'ecosystem-based' field guides for the natural history of a particular ecosystem. There are of course many excellent field guides out there, but this series is meant to look at the full range of organisms within a particular ecosystem, and examine them more specifically in the context of their habitat. Also, I wanted to write books that had a large portion devoted to dealing with the environment and ecology of a particular ecosystem, not just because I find that so interesting, but also because so much of that fascinating information is tucked away in the more technical texts.

Layout of the books

I have tried to use a similar format for different ecosystems. The books usually begin with the basics of the physical environment, such as the climate and geology. This is followed by a look at the ecological relationships, such as pollination and seed dispersal. I then examine the major groups of organisms that are typical of that particular ecosystem, from the simplest plants, through the diversity of invertebrates, to the birds and mammals. The end section briefly looks at the history of humans in relationship to this ecosystem. The books then conclude with a short overview of some of the places where this ecosystem can be enjoyed.

Sections of the book

Much of this book deals with the actual organisms found in the particular ecosystem. Obviously, not every single species of plant and animal can be included in such a book. For many organisms I have dealt with recognizable groups, usually expressed as genera or families. Where I deal with particular species, the print is smaller and abbreviated. I have tended to focus on those species that are more likely to be encountered or are of popular interest. Future editions will have more and more species added. Most of the accompanying photographs are referred to in the adjacent text, and some include information on where the photograph was taken, to give an idea of where the species may best be seen.

The sections devoted to particular subjects vary in length for a few different reasons, including the relative ecological importance within the habitat, the number of individuals, the diversity of species within the group, how much is known about the organism, and the interest and use for the general reader.

The 'scale' of the sections also varies, reflecting both the diversity within the groups and also their place in the evolutionary spectrum. For example, the insects have a relative high diversity of species, so their sections are on a larger scale, and are based mainly on order or family descriptions, while the 'higher' and less diverse birds frequently get species treatments.

A scientific basis

Hopefully, the ecosystem guides also have some scientific integrity. Their underlying scientific foundation is reflected in the way the organisms have been placed into a particular order that roughly reflects their evolutionary relationships.

The text is also referenced, with a full list at the back of the book. The referencing can simply be ignored by those who don't care for, but it is in there for those who would like to source the information. While this is not normal for a popular or field based book, it seemed necessary and fair, especially in light of the many great texts and references out there, and the writers, naturalists and scientists who have worked hard to produce them. It also gives the interested reader a chance to further pursue particular books.

Who the books are for

Despite the underlying scientific base, I hope I have summarized concepts and groups of species to a level that most people can read and understand. One of the main markets of the books is biology student groups and the keener ecotourists that I continue to guide and lecture with. With an increasing population, it may also be useful for interested locals who live in or near the habitat. Primarily, it is hoped the book be of particular use to those who interpret the environment for others, from guides working in the area, to those that own and operate accommodation and information centres in the region. In general, the books are aimed at anyone with a curiosity about natural history. In the end however, I really just wrote a book that I wanted to read, and whenever I had to agonize over any decisions regarding format or what to include in the book, I just went back to that original aim.

Through this book, I hope to contribute to the interpretation of the natural environment and the education of both visitors and locals. This is the most essential yet often forgotten level of any conservation effort. The general public need to be shown and told about these fascinating environments, or they simply won't dedicate their money, time or votes to protect them. I hope it also highlights the work being done by the people that support the very information that is interpreted by the tourism industry, the scientists who do the research.

Thanks

Almost all of the field time, research and writing of these books is by necessity solitary work. However, along the way individuals and companies have helped me with identification, ecology, interpretative guiding, tourism, writing this book, selling this book, suggestions, feedback, or just simply employing me, and they include;
John Chambers (owner of Chambers Rainforest Apartments), Dennis Sinkovich (owner of Crossroads Cafe), Darren Gately (owner of Daintree Hideaway), Stuart Worboys BSc (Botanist, JCU), Dave Phoenix BSc (Tourism manager and author), Ray Andrews (Naturalist Guide), Dr. Heather Zeppel (Lecturer, Tourism, JCU), Dr. Betsy Jackes (Botanist, JCU), Dave Clode BSc (Biologist Guide), Greg Gurnier (Bush Guide), Dr. Andrew Krockenberger (Biologist, JCU), Dr. Darrel Kemp (Biologist, JCU), Mark Steadman BSc (Zoologist Guide), Gary and Debbie Hill (The Adventure Company Australia), Sue and Jodi at Noah Creek Restaurant, Venom Centre Kuranda, Allan, Adrian and the rest of the staff at Daintree Discovery Centre, Jane Moores (Operations Manager) & Matthew Fenn (IT) of ETA, and James Mok (Yes Printing, Sydney). I'd also like to thank many companies and institutions, including The Adventure Company Australia, Silky Oaks Lodge, Chambers Rainforest Apartments, Possum Valley, Daintree Discovery Centre, the Australian Butterfly Sanctuary, Coconut Beach Resort, Lake Barrine Cruises, Paronella Park, Down Under Tours, Coral Princess Cruises, James Cook University and Educational Tours Australia.

Environment

Defining Tropical Rain Forest

There are many colloquial names for 'tropical rain forest'. In Australian colonial times the rain forest was described as the 'scrub' or 'brush'. In the recent past the most commonly used name around the world has been 'jungle', and for many people this probably still covers what we think of as tropical rain forest. In the last few decades, however, the words 'rain forest' have actually replaced the old fashioned 'jungle'. These words "rain forest" are more of a technical term, and were originally used by a German botanist over a hundred years ago; "*tropischer Regenwald*". In contrast to most of the rest of the world, in Australia we often combine the two words "rain forest" into the one word 'rainforest'.

There are many definitions of 'tropical rainforest', even (and most especially) for scientists. These vary around the world, depending on the classification schemes used. One of the most useful of those schemes is the Holdridge pyramid. This is particularly popular in the Americas where it originated, and it's often used to distinguish between different types of 'rain forest' in parts of the Neotropics, especially Central America. Rain forest in this scheme is actually divided into 'moist forest', 'wet forest', and 'rain forest'. It uses the factors of water and temperature, and their variations, to determine the type of vegetation. Within Australia, a system designed by Specht is commonly used. It is based on canopy cover and the height and form of the dominant vegetation. In this classification, tropical rain forest comes under the general vegetation type 'closed forest' (Adam 1997). So, even when we try to scientifically define 'tropical rainforest' we run into difficulties; what a member of the general public in Australia would consider 'tropical rainforest' may be called 'closed forest' by Australian scientists, but would be considered mere 'moist forest' or 'wet forest' by American ecologists. In the end, we must remember that these are all just human constructs for dealing with the natural world.

Ultimately, the best way to define tropical rainforest is simply to describe its characteristics. While many species of flora and fauna are shared between rainforests, most are unique to that particular region, therefore we don't usually recognise tropical rainforest on particular species. However, the different species within these rain forests are under the influence of the same environmental pressures, and thus evolve similar features. These similar characteristics, regardless of geography, can then be used to recognise a tropical rainforest.

Recognizing Tropical Rain Forest

We can use two main types of characteristics to recognize a tropical rainforest; the general structural characteristics of the overall rainforest, and the more specific features of the plants themselves.

General structural features

There are various obvious structural features of the rainforest. For example, it has a multi-layered canopy, which consists of an effectively complete top layer, while below there may be many different levels or strata of plants. The very bottom floor of undisturbed rain forest is relatively open with little undergrowth such as grasses. There is a predominance of vines growing in and around the trees. Epiphytes are another obvious component, with plants literally covering other plants.

Specific features

There are also more specific features of the organisms that live within the rainforest. For example, tropical rainforest leaves are often very similar despite growing on unrelated plants; they are usually leathery, a nice glossy green, and many possess the distinctive 'drip tip'. Many tropical rainforest trees develop features that are rare outside the rainforest; they may flower and fruit on the branches and/or the trunk ('ramiflory' and 'cauliflory), and the roots of many trees spread out as much horizontally as they do vertically, developing into 'buttress' or 'plank roots'.

Many of these general and specific characteristics will be examined in much more detail later in this book.

Distribution

Requirements for Tropical Rainforest

Tropical rainforest requires certain conditions to grow. There is a range of these conditions, including some conditions that must be absent (see 'limiting features' on the next page). At the very basic level, however, there primarily needs to be light and warmth, and moisture.

Light and moisture

The tropics receive a lot of sunlight. This is due to their location on the larger, bulging equatorial region of the globe. There is also little variation in the amount of sunlight during the day throughout the year, with the time of the sun setting and rising not varying much more than about an hour of the same time every day. This consistent sunlight results in constant and reliable high temperature. Thus the temperature does not vary much during the year compared to temperate regions. On the constantly shaded tropical rainforest floor, the temperature varies even less.

And then there is the rain. Despite the big clue in the name '*rainforest*' naive visitors are still often shocked by how much it actually rains in the forest. It is of course an essential requirement for rainforest. And not only does it rain a lot, it needs to rain all the time. Tropical areas have a wet season (some equatorial areas may have two wet seasons), when the rainfall peaks. However, these should be referred to as "*wetter* seasons", as the rainfall needs to also occur in some amounts throughout the year.

Other conditions

There are other secondary conditions that may determine whether a tropical rainforest will develop. One of these is soil type. If the soil is deeper and more fertile, rainforest may able to develop in areas with less rain or lower temperatures that would normally limit its growth. In fact, this is the situation in large parts of the eastern and southern Amazon of South America where tropical rainforest grows (Whitmore 1999). Following on from this, contrary to many popular beliefs, tropical rainforest does not limit itself to poor soils. Rather, it can grow in wet and warm conditions *despite* poor soils. Some of the most diverse and luxuriant rain forests in the world grow on rich soils, such as the rainforests on the volcanic soils in areas of Costa Rica, Indonesia and the Atherton Tablelands in Australia.

'Limiting' features

In addition, there are also certain 'limiting' conditions that cannot occur if tropical rainforest is to develop. The temperature must not drop so low as to allow frost to regularly appear (Winter and Atherton 1987), for this can kill the seedlings of many tropical rainforest plants. The clearing of the rainforest in upland areas often creates open areas that result in frost, thus slowing down the recolonisation of forest in such situations.

A condition known as 'moisture stress' cannot be prolonged if rainforest is to develop. If a true dry season of very little rain lasts for a period of a few months every year, then rainforest will not survive. Depending on various physical environmental factors, including soil and aspect, a 'monsoon forest' (with a different composition of species with dry season adaptations) may grow there, but usually savanna (tropical open woodland and grasslands) will result. The constant moisture means that fires rarely occur naturally in tropical rainforest, except in the unusual situation where there has been several drier years in a row, as happened in the hills behind Cairns in 2003 (as pictured below). Thus, most of the plant species in Australian tropical rainforests are 'fire sensitive' compared to the fire-adapted plants in adjacent savanna habitats.

Distribution of tropical rainforest around the world

Tropical rain forests should thus develop in areas where these conditions are met around the world. These rainforests are, by definition, within the tropics, the region bordered by the tropic of cancer in the Northern Hemisphere and the tropic of Capricorn in the south. In total, they comprise about 7% of the world's surface (Osborne 2000).

Looking at the map below, we can see that there are several distinct tropical rainforest regions in the world. These rainforest regions correlate with the main biogeographical realms (see any biology text for a map of the biogeographical realms). There is Neotropical rain forest (Central America and South America), Afrotropical rain forest (mainly central and western Africa), Oriental rain forest (mainly Southeast Asia) and Australasian rain forest (New Guinea and North-east Australia). As you can see, roughly half of the world's tropical rain forests are in the Neotropics (Whitmore 1999). The smallest major zone comprises the Australasian tropical rain forest. Although this and the Southeast Asian regions are quite close, they are biologically very distinctly different regions of tropical rain forest.

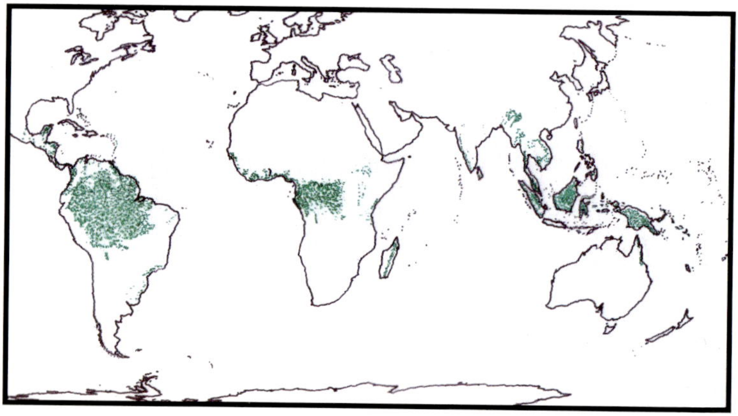

Comparisons to New Guinea and Pacific Island tropical rainforest

The rainforest found on the large landmass of New Guinea (photo, opposite page) and surrounding islands shares many of the same species found in the rainforest of Australia. Both New Guinea and Australia are part of the same continental plate, and thus share the same Gondwanan origin of plants and animals. For much of their history they have been alternatively separated and joined, allowing both isolated speciation and interchange (respectively) of rainforest and it's species. Most of the groups of animals (and indeed many species) are the same, however, the situation with the flora is more complex. New Guinea is closer to the Oriental region and thus many Asian species of plants have made their way across to New Guinea (but have not reached the Australian mainland). For example, the Dipterocarp trees, which dominate the tropical Asian rainforest, but are completely absent from Australia, are found in New Guinea. New Guinea has much more rainforest than the Australian mainland (both in the absolute sense and also as a percentage of land area), higher species diversity, and there are many more different types of rainforest due to the extreme variations in altitude.

The distribution and types of rainforest in Australia

In previous times, rainforest has covered much larger areas of Australia (refer to the "Evolution" chapter later in this book). However, rainforest in Australia is presently an unusual vegetation type in contrast to the rest of the continent. Most of Australia is a very dry country; in many areas, the evaporation rate exceeds the rainfall, and most 'rivers' dry up before they ever reach a sea. Thus, rainforest occupies some 20,000 square kilometres, which is only 0.2% of the entire continent (Goosem 2000).

The extent of tropical rainforest is even less than this, and its distribution is not straightforward due to the other types of rainforests across the continent. Early Australian ecologists divided the basic types of rainforest in Australia up roughly based on distribution and some associated environmental factors, and ended up with 3 main provinces -

Temperate rainforest - cooler rainforest found in Australia in Tasmania and Victoria, and in the higher altitude areas of New South Wales and Queensland.

Subtropical rainforest - including what is collectively known as the 'eastern rainforest', including tracts on the border of northern New South Wales and southern Queensland, such is found in Lamington National Park.

Tropical/monsoon rainforests - there are two main types of rainforest in this group. The 'monsoon forests' have a more open structure than the true tropical rainforest, lower general diversity, and different species composition. It is found in small patches throughout northern Australia, from northern Cape York, across the top end including Kakadu and Darwin, and to the Kimberley region of north-western Australia. The true tropical rainforest as defined earlier is found along the tropical east coast of Australia, mainly north-east Queensland and southern Cape York.

It is also important to note that there is some overlap between these groups due to the correlation between altitude and latitude; that is, the higher in altitude, the more like a higher latitude rainforest it is, due to the temperature dropping as either altitude or latitude values increase. So there are patches of what is really essentially subtropical rainforest in the higher altitude 'tropical' areas of the Atherton Tablelands.

This book deals primarily with this last type of rainforest, the 'true' tropical rainforests of the North Queensland and the rest of Cape York. However, it also covers by default the monsoonal forests, the east coast subtropical rainforests, and even New Guinea (pictured left) and nearby Pacific Islands, because of the large number of plant and animal species shared between these areas.

Distribution of tropical rainforest in Australia

In all, only some 0.1% of the countries' land mass is covered in tropical rainforest. The limited distribution of tropical rainforest in Australia is due to restrictions in various factors such as topography and climate, but in the north-east of the continent, conditions are much more favourable. The presence of mountains so close to the east coast means moisture laden winds coming in off the Pacific Ocean try to rise above the ranges, but they cool and then drop their moisture. This combined with the monsoon winds hitting the tropical coast of Australia results in a high and constant amount of rainfall on the coastal lowlands and ranges of North-east Queensland, creating by far the highest precipitation on the continent. Thus, this area ends up with the largest concentration of tropical rainforest in Australia, some 7,800 square kilometres, and is known as the 'wet tropics'. It sits roughly between 15 and 20 degrees south of the equator, well above the tropic of Capricorn. It has the town of Cairns at its rough geographical centre, and includes the Daintree region, the Bartle Frere and Bellenden Ker ranges, the Mission Beach lowlands, and the Atherton Tablelands.

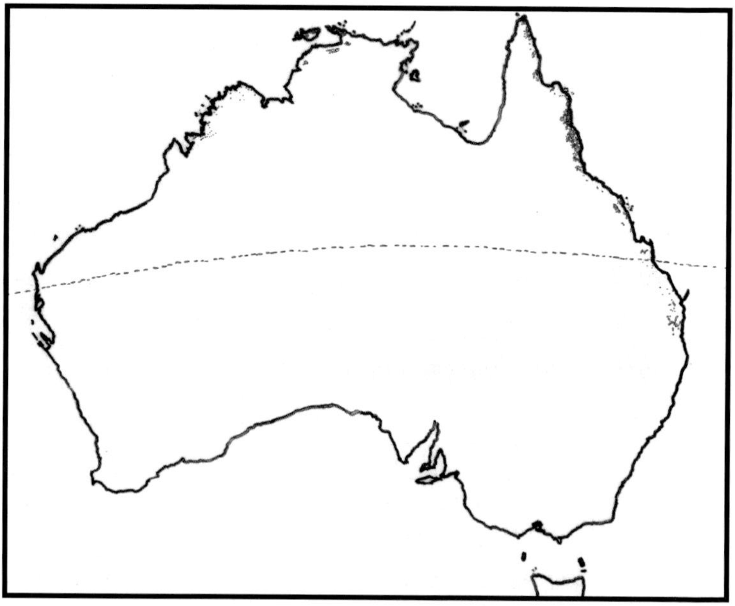

Types of tropical rainforest in Australia

There are further attempts to classify the tropical rainforest in tropical Australia itself into different types. Perhaps the best known and most used system was based on many different factors such as soil, physical structure and dominant plant species, and was divided into a dozen numbered types, with lettered subgroups. The resulting summary was called "Vegetation of the Humid Tropical region of North Queensland" (Tracey 1982/7) and has been used by land managers ever since.

However, to use this system one needs to be able to recognize various environmnental factors and identify species. There are simpler, broader ways to divide the rainforest, and this book will refer to the following basic types of tropical rainforest within the region, based on their geography and altitude.

Lowland Tropical rainforest - This is the 'classic' rainforest, and usually has the highest diversity of plants and animals. In Australia, it is found in places such as Mission Beach, and best expressed in the Daintree region, where there is more lowland rainforest than anywhere else in Australia. Unfortunately much of the other previous areas of lowland rainforest in North Queensland has been cleared.

Mid altitude rainforest of the foothills, ridges and gullies - This is the rainforest of the ridges and gullies behind the Daintree coast, and is best expressed around the foothills and slopes of the Bartle Frere range and surrounds. Because of it's inaccessible nature, much of this rainforest remains, and is protected in areas such as Wooroonooran National Park.

Mid altitude rainforest of the tablelands - Some of the richest rainforest is that found on the Atherton Tablelands. Here, the gently undulating volcanic soil was once was covered in rainforest. Unfortunately, this type of rainforest has been severely cleared, and only small patches remain in many places, such as around the Crater Lakes.

High altitude rainforest - Over about a kilometre in altitude, the rainforest changes substantially in character. This 'tropical rainforest' is in fact much more 'sub-tropical' in character. High altitude rainforest can be seen at Hypipamee Crater and on the peaks of Bartle Frere where at the very top it grades into a **cloud forest.**

Landforms and geology

The very existence of tropical rainforest and other ecosystems in Australia is in part determined by the shape of the land itself.

The coastal plains

Some of the best known, well publicised and most heavily visited rainforest in the wet tropics region is in the coastal lowlands. On a world scale, lowland rainforests are usually considered optimal tropical rainforest. By their very definition the coastal plains are flat, and have therefore been more accessible historically. Thus, much of the coastal lowlands around North Queensland have been cleared for sugar cane and urban areas. While there are remaining patches, for example along the Mission Beach and on Hinchinbrook Island, the largest remaining examples of coastal rainforest are along the Daintree and Bloomfield coasts (pictured below). The shape and size of the coastal land strongly influences the vegetation that grows there. The areas of wider coastal plains where the coastal hills are smaller and the mountains further back, result in freshwater slowing down as it flows from the higher ground, and from the sea allows saltwater to invade further inland. This results in freshwater and mangrove swamps dominating the landscape, for example around Cairns. The narrower coastal areas squeezed between the sea and higher coastal ranges, with their faster flowing water and higher rainfall (dependent on aspect), will often result in tropical rainforest, such as the Daintree coast, where thick rainforest famously grows right down to the ocean (pictured below).

Compared to the older soils of the adjacent foothills, mountains and tablelands, much of the soils underlying the coastal plains are considerably younger. Alluvium soils have been deposited by the flooding and changing of courses of present rivers, such as the Daintree and the Barron. Such soils underlie much of the sugar cane areas around the larger river floodplains. As well, the different sea levels over the last two and a half million years of the Pleistocene Ice Ages have meant higher sea levels have directly dumped material along the present coastal plains. This has resulted in the formation of sand ridges on the coast, such as along the Mission Beach coast, around Port Douglas, around the mouth of the Daintree River, and at Cape Tribulation. Nevertheless, under very wet conditions, rainforest can still grow on these relatively poor, sandy soils.

Rivers of the lowlands

Rivers are obvious to the visitor as they provide so much in the way of recreational life, and they can not be avoided as they have to be traversed while travelling the region. As many small but fast freshwater creeks feed into the coastal lowlands they join and slow to meander as large and murky rivers. The further back the mountains are from the coast, the larger the flat coastal floodplains are, and the more extensive these river systems become. The larger rivers on the more expansive coastal plains are eventually influenced by the tide twice a day, often a considerable way upstream, and thus salt tolerant mangrove trees can be found along their banks for many kilometres. A rise in the land of just a few metres in altitude, up and away from the salt water, can result in other vegetation types developing adjacent to the mangroves, such as Melaleuca swamps and tropical rainforest. These varied vegetation types can be seen in the photo above, represented by different shades of green. Good examples of these rivers are regularly interspersed along the Cape York and North Queensland coast, from the Escape River in northern Cape York, to the Daintree, the Barron River near Cairns, the Mulgrave-Russell system, and the Hull River at Mission Beach.

Rainforest creeks

Further upriver from the coastal lowlands of the coast are smaller rainforest creeks. They are often attractive waterways, flowing with fresh, clear water. This faster flowing, clearer water is relatively nutrient poor, and the abundance of aquatic creek life, such as fish, is poor relative to downstream. However, these freshwater streams provide habitat for many species of creek dwelling frogs that cannot tolerate saltier conditions downriver, and these waterways are lined with a high diversity of rainforest plant species. In North Queensland, many of these rivers are littered with large granite boulders. Accessible examples of such freshwater creeks include Mossman Gorge, Babinda Boulders and Josephine Falls.

The ridges and ranges

Above the coastal plains are the ridges and mountains that provide such a scenic backdrop to much of North Queensland. These landforms are at the northern end of the Great Escarpment, a series of dramatically steep slopes and valleys that occur along almost the entire east coast of the continent. They are associated with the eastern part of the ranges that Australians commonly refer to as the 'Great Dividing Range'.

The ridges and slopes of the edge of the escarpment around North Queensland are composed mainly of granite (see section below) and metamorphic rocks. The latter were originally sediments laid down from about 400 million years ago, in the seas of the Silurian and Devonian periods. As more and more sediments were piled on, the lower layers were compressed and metamorphosed into slates and greywackes. Later compression and buckling resulted in a mixture of metamorphic and sedimentary rock in what is now called the 'Hodgkinson formation'. Due to it's rugged nature, most of the escarpment is underdeveloped, and this is where much of the remaining protected tropical rainforest is found. The roadways from the coast to the tablelands go through these landforms, such on the Palmerston, Gilles, Kuranda and Mt. Molloy highways, and the rugged topography allows for some spectacular views.

The high granite peaks

The very tops of the ridges and the mountains are of a different geology to much of the surrounding slopes and gullies. Many of the higher mountains and peaks are composed of granite. Much of this probably formed in the Carboniferous period. This rock begins as molten material, magma, that remains underground and thus cools very slowly into a coarse grained hard rock. It often 'intrudes' into surrounding and higher layers. The hardness of granite means that if there are softer layers above that are eroded away, it's resistance results in extruding mountain tops and boulders in the landscape. And thus in many parts of North Queensland the softer surrounding metamorphics were worn away and left massive granite block inselbergs, such as atop Thornton Peak, Bartle Frere and the ridges along the backbone of Hinchinbrook Island. Further, as granite is made up of coarse crystals, the water tends to percolate along the edges of these large crystals, and weathers along developing lines of weakness. This can result in large fractures around blocks that then may break up into the large rounded boulders that are often associated with granite formations, such as those that can be seen in and around the Mossman River. Even more dramatic examples of this weathering of granite into blocks can be seen at Black Mountain National Park just south of Cooktown, and in Cape Melville National Park at Princess Charlotte Bay. The resulting granite-derived soils downslope and downriver of these parts of the can be fairly acidic and nutrient poor.

Volcanoes and the tablelands

After driving along the highways through and over the steep Great Escarpment, one comes quite dramatically and suddenly to the rolling plains of the Atherton Tablelands. These undulating landscapes vary between about 600 metres and almost a kilometre in altitude. These tablelands were the site of some of Australia's most recent phase of volcanic activity, beginning some 3 million years ago to perhaps only 10,000 years ago. Over 50 eruptive features have been found over the Tablelands (Ferrett 2005), and they include a diverse and accessible range of features. The biggest volcanoes are the hardest to detect, being low but large 'shield volcanoes' and include Hallorans Hill and Bones Knob near Atherton.

In the undulating central tablelands around Yungaburra are Mt. Quincan and the 'seven sisters' (pictured below), a series of 'cinder cone' volcanoes. They form a rough line, suggesting they formed along a fissure or line of weakness (Ferrett 2005). The dramatic Hypipamee Crater (pictured above) is probably the best example of a 'diatreme' in Australia. It begins when underground water is heated from below, builds up pressure, cracking the overlying rock, and when it is finaly released, the super heated water expands suddenly in a massive steam explosion. It was powerful enough the blow through the existing granite, leaving a 55 metre crater, with water at the bottom that goes to a depth of over 80 metres (Ferrett 2005). Such diatremes may develop into 'maars' when it is molten rock that has heated the water causing the steam explosion. Examples of these include Lake Eacham and Lake Barrine, as well as Euramoo near Tinaroo, and Bromfield Swamp. The susequent crater eventually fills up with rainwater. The crater lakes were probably formed between 20,000 and 10,000 years ago, well within the time people inhabited the area. Many of the waterfalls, such as the Malanda, Mungalli and Millaa Millaa Falls, go over previous lava flows which have resulted in formations known as 'columnar basalt'. Millaa Millaa falls goes over two different patterns of basalt columns, the result of two separate lava flows. Lava flows that spilled into the area from volcanoes have resulted in the pleasant undulating hillsides that are known as the 'tablelands'. This has also resulted in rich basaltic soils. Such soil tends to be red in the tropics, rather than the dark brown or black in other areas, and it can be clearly seen in the road cuttings around the tablelands. The relatively flat land and rich volcanic soils has resulted in massive clearing of the tablelands, and the land being used for a huge range of agricultural activities. As one goes further west and inland, the altitude slowly drops, the soils become less rich, and the rainfall and the forest fades away.

Climate

The seasons

Australians living in the tropics generally recognize two main seasons; the 'wet' and the 'dry'. These two seasons are based mainly on the amount of rainfall. As part of tropical monsoonal Australia, the "Wet Tropics" experiences a distinct 'wet season', which occurs in the southern hemisphere summer. The southern winter is much drier in most of tropical northern Australia, but due to the winds and local topography, the north-east Queensland region still experiences sufficient rainfall in the so called 'dry season' to sustain rainforest. In contrast, those people that relied on their environment the most, the indigenous Australians, recognized many distinct seasons throughout the year. The number and length of the seasons varied, depending on the group and where they lived in tropical Australia (e.g.; Morris 1999).

Sunlight and temperatures

Temperatures on the coastal plains of north-east Queensland generally don't vary by more than 15 degrees Celsius throughout the year, with temperatures in the early 30's being the average maximum in the summer season and high teens being the average minimums in the winter season. However, the high humidity of the wet season make the heat of the summer seem much more oppressive to many first time visitors to the tropics.

Temperatures get much lower in the higher altitude parts of the Atherton Tablelands, and much of these areas are considered to have rainforest which is more 'sub-tropical' in character and species composition. In areas of the tablelands where the forest has been cleared, they may get frost in the open fields.

Precipitation

The "Wet Tropics" are, as the name suggests, the wettest part of Australia. The centre of the region is Cairns, which receives an annual average of just over 2 metres a year (Tracey 1987). In the year 2000 it received over 3 metres, the highest in it's recorded history (Cairns Post, December 23, 2000). Some areas of the Wet Tropics, such as Mt. Bellenden Ker, have been recorded to receive an average of over 8 metres a year (Tracey 1987), with the heaviest recording so far being almost 11 metres. In one 24 hour period 1140 cm fell; over a metre of rainfall. That is on average more than Mareeba, (and in fact most of Australia), gets in a year!

For rainforest to develop there needs to be sufficient rainfall throughout the year; however there is definitely seasonal variation. There is a peak of rainfall, with most rain in this region falling in the summer or the 'wet' season, from December to February, (although the heaviest rainfall is actually in early 'autumn' in March, Tracey 1987). August and September are usually the driest times with a monthly average less than half a metre (Tracey 1987). There is often fine weather right up until Christmas.

Orographic effect

With it's mountainous coastline, one of the most influential factors in the north-east Queensland area is that of the orographic effect. Moist air comes in off the ocean onto the coast, and when it hits a barrier, such as a coastal mountain range, it is forced to rise. As the clouds rise they cool, condense and lose their moisture. This continual coming in of warm moist air results in high precipitation along coasts with mountains.

In fact, the mountains don't need to be that high; a rise of only 50 metres can increase rainfall precipitation as much as 25% to 50%. However, the higher the mountains and the closer they are to the coast, the more significant the effect. In such areas, the result is a high rainfall ecosystem, such as rainforest. This is important in determining much about the 'wet tropics' of north east Queensland.

Cloud stripping

Direct rainfall is not the only type of precipitation. When mist and clouds pass over the valleys and peaks, moisture condenses on the plants; a phenomenon known as 'cloud-stripping'. Recent research at James Cook University and CSIRO has suggested that this effect is much more significant than previously realized; at some times of the year, such cloud stripping may be collecting up to 40% more moisture than is measured by traditional rain gauges.

Disturbances

Tropical cyclones

Tropical Cyclones are huge water-borne storms that often do great damage to the rainforest and surrounding areas. There have different names in different parts of the world: typhoons in Asia, hurricanes in the Americas and cyclones in Australasia. Due to the spin of the Earth and the corresponding Coriolis Effect, they develop their distinctive spiral patterns and rotate in a different way in different hemispheres (clockwise in the southern hemisphere); but they are essentially the same type of phenomenon. Many of the rainforests in the central equatorial areas of the world are in fact not greatly affected by cyclones, as the incidence of cyclones is much higher towards the edges of the tropics. Tropical Australia lies in this zone and is thus well known for its regular cyclones. It is likely that few areas of tropical rainforest would not be affected by major cyclone damage every forty years (Adam 1997). Many areas lowland areas of north-east Queensland, such as around Mission Beach and the Daintree, are hit much more regularly by large tropical cyclones, and so develop quite distinctive 'cyclone scrubs' along their coasts with particular growth forms and typical species, such as Cyclone Vine (*Merrima peltata*). The most obvious effects of cyclones are the huge winds and damage caused by tearing apart of buildings and flying objects. However, it is the constant rain and subsequent flooding, especially if is occurring at the same time as king tides, that causes much of the damage and also problems in transport and communication.

There have been several bad tropical cyclones in recent decades. With the increase in urban population and property in North Queensland, the study of tropical cyclones has become essential. At the end of last century, Cyclone Rona hit the Cairns area and isolated the northern beach suburbs with the accompanying floods. Cyclone Steve hit Cairns in 2000. The most significant cyclone of recent years has been Cyclone Larry. This hit the coast south of Cairns in the late wet season of 2006, with the Innisfail and Mission Beach areas most severely affected (pictured below), but even many parts of the Atherton Tablelands, including iconic spots such as Lake Eacham, had their rainforest knocked down. Cyclone Monica immediately followed, dumping large amounts of rain in the region.

Evolution
& adaptation

Evolutionary history

Over the last few decades information about the evolutionary history of the Australian tropical rainforest has slowly been revealed, and with that it's importance in both an Australian and international context.

The history of the Australian continent

It is now widely accepted that the positions of the continents, and their countries, has changed over time. This idea is called 'continental drift' and is explained by the mechanism of plate tectonics (e.g.; White 1998). It is thought that much of the present land on our planet had it's origins in various super-continents. The last super-continent was called 'Pangaea' and it split into two continents called 'Laurasia' and 'Gondwana'.

Laurasia was composed of much of what we now call Europe and North America, while Gondwana comprised major sections of the so-called 'southern continents' such as South America, much of Africa, Australia, New Guinea, Antarctica and India. Gondwana evolved its own distinctive flora and fauna. This connection has been used to suggest the possible reason of the distribution of certain groups, such as Proteaceae flowers, ratites (including the cassowary, pictured right), parrots, and others that are currently more common in the Southern Hemisphere continents. Eventually, Gondwana broke up into pieces and Australia began to drift in isolation towards South-east Asia. It evolved it's own unique flora and fauna from Gondwanan stock over millions of years. At this stage Australia appears to have been covered in much rainforest (White 1998), and based on the occurrence of more primitive species in higher altitude rainforests today, scientists guess that much of this rainforest was subtropical in character (Adam 1997). This is also the time when Australia's unique animals groups, such as kangaroos and koalas, evolved.

This isolation has gradually been coming to an end as Australia nears Asia at a rate of about 7 centimetres a year (White 1998). Many plants and animals have begun to cross over from Asia into Australia, and vice versa. However, this continuing process is a relatively geologically recent event, and the two regions of rainforest in Australia and South-east Asia, although now very close, have very different origins, and thus very different flora and fauna.

A unique and ancient tropical rain forest

Not so long ago Australian ecologists thought that the rain forest in Australia was essentially Asian (e.g. Clark 1979). It was referred to as the 'Indo-Malayan element' of our continent and was thought of as 'invasive'. The rainforests were not considered truly 'Australian elements', as were the *Eucalyptus* woodlands and *Acacia* shrub lands. Eventually, scientists began to realize that our rainforests were a distinctly Australian element. Palaeontologists began to unearth rainforest plants and animals, thus demonstrating that this habitat had long existed on this continent. The small fragments of rainforest left on the continent today are all that is remains of rainforest that previously covered huge tracts of the Australian continent. As scientists examined the plants and animals of these ancient forests and studied their phylogenetic relationships, it appeared they were the ancestors of much of the flora and fauna found in drier more widespread environments today. Thus, it appeared that much of the flora and fauna in the drier open habitats of present day Australia had actually evolved from original rain forest stock (Winter and Atherton 1987).

And so many of the animals and plants that exist in the rain forest today may be similar, or related, to these ancestors. For example, the open forest kangaroos we are familiar with today have evolved from a lineage that began with something like the Musky Rat Kangaroo that is found only in North Queensland tropical rain forests today (pictured below). It is also believed there are plants in these same rain forests that represent something akin to the ancestral Eucalyptus trees. Thus, experiencing these rain forests enable us to, in a fashion, go back in time and observe Australia's evolutionary past. So over the last few decades our perception of our own rainforests has turned upside down. The rainforest changed from a small collection of recent invaders from Asia that didn't really belong here, to the 'original Australians' from which many of our species probably evolved. The vital evolutionary significance of our rainforest forms much of the basis for their proposal and acceptance into World Heritage status.

Further, the rainforests of Australia may also be historically important in a more international context. The rainforests of north-eastern Australia and other areas of the south-western Pacific are interpreted by some scientists as the oldest in the world, their evidence provided by what appears to be the highest known concentration of primitive flowering plant families (Adam 1997). This is not to say that flowering plants necessarily evolved here, but that they happen to survive here in 'refugia' for millions of years. Certain protected gullies of the region must have avoided the worst of the arid and cold phases caused by glacial peaks, sea level rises during interglacials, regular and widespread destruction by volcanic activity, and general large geological change (Nott 2005). The concept of the 'oldest rainforest in the world' is fairly subjective however, and used loosely by marketing and environmental organizations, and the 'western Pacific' is a large and varied region; thus for example, Malaysia still promotes their forest as 'the oldest rain forest in the world' (e.g.; Uchida 1996), even though it is very different in species composition.

Adaptations

Light, colour and vision

Many first time visitors to the rainforest expect to see a riot of bright colours; but the forest is much more subtle than that. There are in fact many shades of colour out there, especially green. There *are* bright colours, but you have to search them out. New leaves and fruits provide much of the stationary colour in the Australian tropical rainforest. Most of these colours are provided by pigments, and as attractive as they are to us humans, they have evolved for practical purposes. Many of the shades of green in leaves are for the day to day purpose of energy production through photosynthesis, while other brighter colours in new growth have other purposes (see section on "Leaves" later in this book). Brighter colours in flowers and fruits have primarily evolved for attracting animals; to eat nectar and thus pollinate the flower, and to eat the fruit and thus disperse the seed. In fact, as humans we only see a certain part of the light spectrum, and we are not picking up all the colours. Many species of flowers have colours at the ultraviolet end of the spectrum specifically to attract insects that can see in these wavelengths.

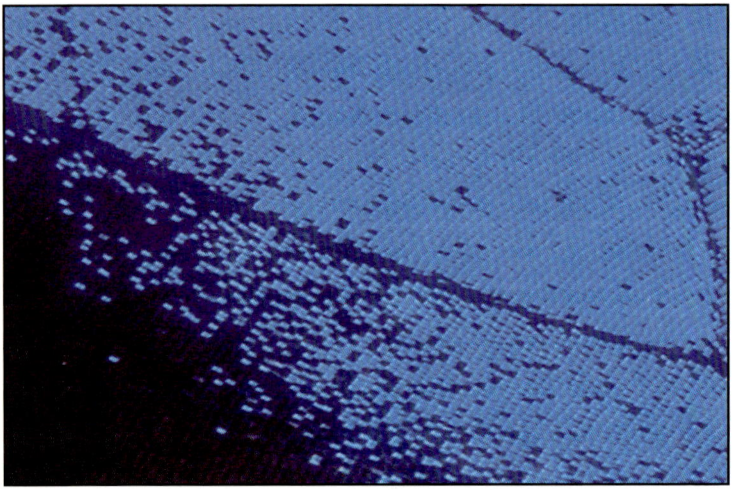

Other colours are not provided by pigments but are known as 'structural colours'. They are caused by special layers in the upper cells interfering with the light, and causing the brilliant and changing iridescent colours that we see on rainforest animals such as the Ulysses Butterfly and the Riflebird, a bird of paradise. The bright blue of the Ulysses butterfly wings are spectacular on a sunny day, for the brighter the light, the more dazzling the colour. This allows the animal some form of control over it's colour. The bright blue of the Ulysses disappears and the animal virtually vanishes when it lands and closes it's wings. The Birds of Paradise can position themselves in a sun lit area and fluff out their feathers to take full advantage of their refraction based colours when displaying. Structural colours are very rare in plants, although there is an exception provided by the Australian tropical rainforest; the bright blue fruits of the Blue Quandong.

Other animals use their colours to blend in to their background. Some techniques are easy to understand; if the organism is dull in colour, it disappears into it's background. A good example is the Northern Barred Frog, using it's leaf like browns to blend into the forest floor (pictured below, Hypipamee Crater). However, other techniques are counter intuitive. Ironically, brighter colours might help animals blend in. Bright colours can be used in a patchwork pattern to break up the outline of the animal. Examples are colourful parrots such as Rainbow Lorikeets which can be surprisingly hard to see in the canopy of the forest.

Sound, listening and calls

The rainforest is often quieter than many people expect. Depending on the light levels and the time of the year, the dawn chorus of birds is loud, while at other times it may be fairly subtle. The thickness of the vegetation muffles sounds, and low frequency sounds and single, sharp calls travel better in this sort of environment. These are the sorts of sounds that are more common in many bird calls, especially those that frequent the deep interior of the forest, such as the ground dwelling Chowchillas with their loud 'star wars laser gun' calls, and the Cassowary (pictured right) with it's low frequency rumble.

Evolutionary trends

Convergent evolution

Most of the plants and animals found in the different rainforests around the world are different species, and in some cases, they are not even closely related. Yet they have evolved in similar habitats under similar pressures and thus often end up looking quite similar. This is convergent evolution. An example relevant to the Australian tropical rainforest includes a comparison between the Green Python of Cape York and the Emerald Boa of South America; both are bright green as adults, both have bright yellow juveniles, and both have a distinctive resting posture looped on a tree branch (pictured below; boa on the left, python on the right). Boas and pythons are related groups, but convergent evolution also occurs in animals that are not closely related at all. For example, the sugar gliders of the Australian rainforests and the flying squirrels of the south-east Asian rainforests are both small mammals with fluffy tails, large ears, and membranes that enable them to glide between trees in the rainforest canopy. However, squirrels are eutherian or placental mammals and gliders are marsupials.

Co-evolution

Part of the diversity of the tropical rainforest is attributed to the co-evolution of groups of organisms. This can occur in several different ways. For example, animals and plants can evolve in a cooperation. The most obvious case in the tropical rainforest is the co-evolution between flowers and pollinators. To encourage the most efficient disperser of their pollen, certain plants have evolved flowers to attract specific types of animals (refer to later section on pollination).

Co-evolution can also result from the competition between groups of organisms over time. A classic case of this is the 'arms race' between plants and the caterpillars that feed on them. If a new plant develops a certain defensive toxin to avoid being eaten, eventually surviving caterpillars evolve to feed on the new plant, followed again by new plants evolving different toxins to avoid being eaten by those new caterpillars, and so on. This continues, with new species of plants evolving new defences and new herbivores evolving to feed on them.

Structure and forms

The general structure

The tropical rainforest consists of a multi-layered canopy. The main canopy can consist of an undulating 'sea' of tree crowns with heights ranging from 15 to 45 metres in lowland tropical rain forest (Whitmore 1999), with the cyclone exposed and nutrient poorer ridge top and coastal rainforest tending to be shorter, and the protected and nutrient sink valley and gully forest growing taller. Every now and then, taller trees poke out above all else. These trees are known as 'emergents' (Johns 1999). In the swampier lowlands of the tropical Australia these emergents are commonly tall, straight palms such as Black Palms and Alexandra Palms.

Below the main canopy may be many different levels or strata of trees. These catch any light that manages to get through the first layer. The result is an overlap of many different levels of trees. Looking up from below, the canopy appears closed. Only small gaps reveal the sky above. This then results in very little of the light reaching the floor of the tropical rainforest.

Light and the floor of the rainforest

Measurements in some rainforests have shown that as little as 2% of the sunlight may reach the forest floor. This means that the very bottom floor of the rainforest has little undergrowth, such as grasses, and thus undisturbed rainforest may have a relatively open understorey. The plants that do grow down here in the comparative darkness often have relatively huge leaves to catch what is left of the sunlight. In the lowland forests they belong to the giant herbs, such as native bananas and gingers.

The edge of the rainforest

The edge of the forest is a continually growing mass of vegetation, and is often thick and impenetrable. It is essentially the canopy growing down to the ground. It is a good place to observe flowers and wildlife of the canopy that may not be otherwise accessible. These edges can be natural, such as along waterways, or human induced, such as along roads. The structure and edge of the tropical rainforest can be appreciated with views from the Palmerston and Kuranda highways, or on boats along the upper Daintree River or on Lake Barrine.

Growth of the rainforest

Disturbance and gap creation in the tropical rainforest

Within the complex multiple layers of the canopy there are always different stages of growth, depending on local patches of constant disturbances. These gaps contribute to the overall physical and species diversity of the rainforest.

Regularly, a gap may be created in the canopy. On the coastal lowlands, disturbance is often the result of storms and tropical cyclones. Huge holes in the canopy grow over in a matter of months. Areas along the coast that are repeatedly battered over the years develop a distinctive, recognizable structure and end up with a particular suite of species of plants smothering them, such as the climbing rattan palms and the appropriately named 'Cyclone Vine'. Or there can be a single tree fall, perhaps from a lightening strike, or simply the natural end of a life of a relatively fast growing tropical tree. This falling tree often brings down a mass of other plants through the tangled network of connected vines and epiphytes.

Such disturbance and gaps allow new light through. Other surrounding plants are quick to take advantage of this. Before long there is vigorous growth and an 'explosion' of tangled plants, such as the notorious 'wait-a-whiles'.

Pioneer and climax species

When plants do grow back in the gaps or on the edge of the forest, how they grow and respond differently to light and disturbance determines an important way of categorizing plants. At one end of the spectrum there are the 'pioneer species' and at the other end are the 'climax' species, with various grades in between.

The pioneer species are those particular species that grow aggressively and quickly on the edge of the forest or to fill up a gap. Their seeds will only germinate in an area where a gap will allow full sunlight to the ground for some part of the day, and the seedlings are not found under a full rainforest canopy (Adam 1997). These plants tend to live fast; growing quickly and dying young, living only 15 to 20 years (Adam 1997). There are many less pioneer species than the climax species, but the pioneers tend to be geographically widespread, with the same species acting as pioneers in many different areas of rainforest. Being on the edge of the forest, they are usually the better known species. Typically, pioneer species produce many seeds, and they are often small so they may be wind-dispersed from their exposed position on the edge of the forest. The species that are pioneers are not necessarily related, although it so happens that many are in the Australian rainforest belong to the family Euphorbiaceae, such as the Bleeding hearts and Macaranga. The latter genus has many pioneer species throughout tropical Australia, New Guinea and tropical Asia, and some can grow 8 metres in a year.

Pioneers eventually get taken over by slower growing, harder wooded trees. These are the climax or non-pioneer species. The seeds of non-pioneer species are capable of germinating underneath the canopy, however, they can also grow in gaps, but they are usually initially taken over by the faster growing pioneers. However, they may eventually grow up into the rainforest giants, and provide the ultimate canopy. In contrast to the small, but widespread, number of pioneer species, the non pioneer or climax species are the majority of species in the tropical rainforest, and many may be endemic to smaller areas (Adam 1997).

Seedlings

Seedlings have to be dropped away from the parent tree to avoid competition with the parent tree. The distance between the parent tree and the fallen dispersed seed is called the 'seed shadow'. Many rainforest trees seem to have quite a short seed shadow. If animals do not disperse the fruit, or wind dispersed seeds do not fall far from the parent tree, a blanket of seedlings may grow near the tree.

These may grow to a few inches and then slow their growth to such an extent that there appears to be a permanent carpet of miniature plants frozen in growth. Many rainforest seeds exude a pectin-like substance which absorbs water and forms a gel; this dries up then glues together leaf litter around the seedling, creating a little micro-environment (Milton in Nadolny 1999). These seedlings may in fact retain their small stature for many years, even decades. They are biding their time, just waiting for another gap in the rainforest, at which time they will enjoy a sudden burst of growth.

Trees

First time visitors are often surprised that the trees in the Australian tropical rainforests are not taller. There are several reasons for this. It has been found that in general, tropical rainforest in the central equatorial regions of the planet where cyclone incidence is low, are taller, whereas the rainforest on the edges of the equatorial regions where the cyclone rate is much higher, (such as northern Australia) are shorter (Primack and Corelett 2005).

Unfortunately for the casual visitor, it is also likely that most of the easily accessible tropical rainforest in Australia was utilized by loggers for their timber. Apart from some particular protected individuals, such as the twin Kauri Pines at Lake Barrine (pictured right), to see the larger rainforest specimens requires walking deeper in the larger protected areas, for example, around the valleys of Bartle Frere.

In general, the tallest trees in tropical forests are shorter than their counterparts in cooler temperate forests. In North America, much taller trees are found in the cooler north-western forests. And in Australia, the biggest flowering trees in the world are Eucalypts in the schlerophyll forests in the south-east and south-west of the continent.

Growth and age

Age can be difficult to determine in tropical trees for several reasons, including the dating techniques used, and because of differing growth rates. Due to the comparatively good conditions, the average tree in the tropics can potentially grow for much of the year. Thus, the tree rings that are used to traditionally age trees in temperate trees where the growth rates vary greatly during the change in season, usually don't correlate with a determined age in most tropical trees (Adam 1997). Rather, the direct dating of trees is done with radio carbon dating on wood from the heart of the trunk base, and this technique has revealed several very old individual rainforest trees in tropical rainforests around the world (Turner 2001). Trying to predict age by comparative growth rates and size is complicated by the fact that the rate of growth of tropical rainforest trees changes during its lifetime. While growth can occur all year round due to favourable conditions, such as seasonal predictability, steady high temperatures and enough moisture, studies have shown that the limiting factor to many trees in tropical forests is light (Turner 2001). While still a seedling and with minimal access to light, certain species may grow very slowly, if at all. But when they do have access to light, tropical rainforest plants may grow quickly, reaching impressive size in very short time. However, once they reach a certain age and are part of the mature forest, with relatively low light levels and limited space, they may slow down, stop growing or even shrink slightly (Turner 2001). Thus, rainforest trees have very different growth rates that don't depend so much on the time of the year, but rather on the stage of their life and surrounding environmental conditions.

Root systems

The roots of tropical rainforest trees have interesting variations. In many trees, the roots will spread as much horizontally as they do vertically. Erosion by running water may eventually reveal this complex network just underneath the ground, and walkers in the forest find themselves constantly tripping over these exposed root networks. Their intertwined nature means that when a single tree is hit by lightening, it can effect other plants in contact around it. If the tree is knocked over in a storm, the shallow and horizontal root system may lift up much of the surrounding topsoil with it.

Many trees develop large flanges above the ground that flank the trunk on either side. These are called 'buttress' or 'plank roots' and give the tree the appearance of a rocket. Although some species are capable of developing them, and some are not, the development of buttress roots is a reflection of environmental pressures and not an indicator of species. Reasons suggested for such development include support and stability, allowance of air exchange in water-logged soils, and the capture of leaf litter. In the latter, the roots act as a nutrient 'sink', collecting fallen debris for breakdown. These areas are then used by sheltering animals, and thus the tree then collects further nutrients via the animal's waste products

Cauliflory

The flowering on the branches and/or the trunk of a tree is called 'ramiflory' and 'cauliflory' respectively, and a tree that fruits from the trunk is known as 'caulicarpic' (Richards 1998). Early explorers (and even visitors today) have mistaken them for a parasite on the plant (Whitmore 1999). They can make for a spectacular show in the tropical rainforests, and are relatively rare in other habitats. This form of flowering and fruiting has likely evolved to allow easier access for specialized pollination and seed dispersal by shade-loving insects and other understorey specific animals, such as certain species of bats (Whitmore 1999).

Leaves

The canopy of the rainforest is usually closed, with plants retaining much of their leaf cover throughout the year. Fully deciduous trees are uncommon and may only account for a few percent of the tree species in the tropical rainforest.

Shape and size

Rainforest leaves themselves share certain morphological features that are adaptations to the environmental pressures that they all have in common. They are usually leathery and glossy green. Up to 90% may possess the distinctive 'drip tip' (Johns 1999). This is an extension of the end of the leaf, generally believed to have the effect of running the excessive moisture off the leaf (pictured below).

Leaf size in the best developed, lowland tropical rainforest is relatively large, being generally described as 'mesophyll' (Whitmore 1999) or even larger. Leaf size decreases as altitude increases and temperature decreases. Within the same area, and even on the same plant, leaf size varies depending on its position in the canopy, with leaves generally being larger in the lower and darker depths of the forest, and smaller in the higher and lighter areas.

Leaf colour

In the Australian tropical rainforest, much of the splashes of colour are provided not by flowers, but by leaves. Leaves can be yellow, pink, red, purple or almost black. Unlike in temperate regions, many of these colours are not of leaves about to be dropped, but are often flushes of new growth. As the fresh rains and rising temperatures signal the beginning of the summer wet season, there may be subtle explosions of new leaf growth. Many plants in the family Myrtaceae, such as the 'Pendas' and the *Syzygium* have bright red and pink new leaves. Yellow leaves are usually leaves with undeveloped chlorophyll. However, red or pink leaves are coloured by a pigment called 'anthocyanin'. This pigment reflects red light, which is normally used in photosynthesis. It is commonly found in fruits and flowers, and is responsible for the red colour in beetroot (Martin 1990).

These bright colours may occur for several reasons. The presence of the pigment anthocyanin may be to protect the developing chlorophyll inside the new leaves from too much light, or high energy, damaging light such as ultraviolet. It has also been suggested that these colourful pigments may contain fungicides and other poisons (Breeden 1999). Another idea put forward involves the fact that many herbivores prefer soft, fresh growth to the older, tougher leaves and that the brighter colours of the fresh growth may not be recognized as leaves by herbivores, or serve as a warning that the new leaves are toxic or distasteful (Farrant 1999).

Plants growing on other plants

In the relatively wet and warm conditions of the tropical rainforest, the main struggle is for light itself. Thus in the crowded rainforest, plants often grow on anything; rocks, other plants, and even human made objects. The name 'climber' is a fairly vague term that includes plants that grow on and over other plants and other objects. They are defined by their common method of climbing other plants to reach sunlight without using their own support, at least initially. Generally, the greener shorter-lived climbers are known as 'vines', and the longer-lived woody vines are known as 'lianas' although this terminology varies with language and geography and is not strictly scientific. Plants that grow atop other plants are usually referred to as 'epiphytes', with plants that grow on rocks being called 'lithophytes'.

Terms like 'climbers and 'epiphytes' are descriptive ecological terms. Therefore, they do not imply relatedness, but the evolution of certain features from the same environmental pressures. Different species of climbers belong to a huge range of different families, in fact even different divisions. The more obvious epiphytes in the Australian tropical rainforest are ferns, while orchids are very abundant, but not as commonly noticed. There are no native bromeliads in the Australian tropical rainforest; and thus none of the so-called 'air plants' of the Neotropical forests.

Climbers and epiphytes are some of the most obvious aspects of this ecosystem, indeed their existence helps define the 'tropical rainforest' itself. In some classification systems in Australia, certain rainforests are known as 'vine forests' (e.g.; Webb and Tracey 1987). It is not uncommon to see a large number of both species and individuals of climbers crowded on the one tree. In some areas, there may be almost as many climber species or individuals as there are trees. Once secure, the lianas may develop large woody stems that are, to use a jungle cliché, 'as big as a man's thigh'. There is a record of one liana being 60cm in diameter. The 'rattan palms' of Asia and Australia develop the longest stems of any plant in the world, with records of stems over 500 feet (Attenborough 1995). Barring collapse, some of these woody climbers may be as long-lived as many of the trees of the forest. While vines are found all over the world, over 90% of species are found in the tropics.

Biology of epiphytes

Visitors often see these growths in the canopy and may assume they are a natural growth of the tree, or parasites. They may even mistake them for constructed bird's nests. However, epiphytes are simply plants that grow atop other plants. They, like pretty much all other plants, use sunlight for photosynthesis. In the thick vegetation of the jungle, little light reaches the lower levels, so epiphytes are 'getting a leg up' by growing up and on top of other trees.

Many epiphytes include plants that have small seeds. This enables easy dispersal through the air and up into the canopy to find a place to grow from. Orchids have hard capsules that open and release hundreds of tiny seeds. Ferns are even lighter; the sporophytes seen underneath the frond (pictured right) release thousands of tiny dust like spore that, with just a whiff of air, can float up into the canopy to settle on a tree.

Ecology of epiphytes

Epiphytes may need only a small imperfection on the side of the tree or branch to begin growing. Most rainforest trees have these nooks and crannies, and it is a challenge to find a rainforest tree with no other plants growing all over it. However, certain species have adaptations that may have evolved to discourage such growth. All around the tropical rainforests of the world, many species of palms can avoid epiphytic growth. Palms grow from a central point in the middle of the tree and upward, and as the new growth continues, the previous leaves and bases fall off, taking any epiphytic growth with it. Some other local examples include the Kauri Pines and several of the Satin-ashs with their flaky bark that peels off and presumably takes potential epiphytes with it as it falls off.

Many epiphytes can grow to a size where they completely surround and are larger in circumference than the tree that supports them. As they are not parasites and are not taking anything away from their host, it is in fact it is in their best interests not to damage their host so as they continue to retain a nice high rise living. However, the enthusiastic growth of epiphytes may eventually damage a tree; large epiphytes have been measured at half a ton in weight and after rain these water soaked plants may become so heavy that their weight brings down the branches they are sitting upon, and may even contribute to the fall of the entire tree.

Growth patterns of climbers

Climbers in the rainforest create some interesting patterns and formations. Many of these can be explained by simple biology. When lianas seem to hang in the air like an Indian rope trick, they create the question of how they climbed up through empty space. The answer of course is that they didn't; when younger, these climbers grew up trees or through a canopy which has since disappeared through death and/or collapse, leaving the climber hanging.

Some of these climbers hang in a pleasing spiral pattern (right). These can be called 'ghost branches', as when the climber was younger it grew around a branch or tree that has since disappeared; but the climber retains its spiral growth pattern.

There are many misconceptions about the spiral growth of vines around trunks. In early studies of climbers in higher latitudes it was thought the direction of growth was to take advantage of the sun (Whitten *et al.* 1987); however, as most tropical rainforest, (and therefore most climbers), are found in the equatorial regions within the tropics, the overhead aspect of the sun negates any particular spiralling. Later studies have found that a higher percentage of climber species happen to have a genetic disposition to anti-clockwise growth, just as the majority of humans are right-handed (Martin 1992).

Different species of climbers have developed different parts of their anatomy to climb up. Which part of the plant is used for this purpose is in a large part determined by what group of related plants it has evolved from. The vigorously growing Supplejack (*Flagellaria indica*) climbs by using its leaves; the tips are elongated and curl around to catch onto the surrounding vegetation.

Vines in the wine family Vitaceae have tendrils that curl out from the main stem opposite their leaves (Jackes 1999). The wait-a-while palms (found in Asia and Australasia) have developed their tendril (which normally carries the flowers and fruits) into a climbing appendage with recurved hooks (pictured right); any sort of disturbance, such as a slight wind, may end with the light tendril hooking onto nearby vegetation and thus giving the plant a foothold into the surroundings.

Ecology

Herbivory

The biggest areas of ecological interaction involve the relationships between plants and animals. One of the most obvious of these is the act of the animal eating part or all of the plant. Some of this feeding is actually part of the plants' 'strategy'; for example, animals feed on the nectar produced by flowers, or on the fleshy fruits grown around seeds.

However, many animals actually consume the plant itself. The eating of plants by animals is called 'herbivory'. When the animal munches on the plants leaves, it is taking away parts of the plants apparatus for energy production, when the animal digs up and feeds on the plants roots or tubers, it is taking energy stores.

Herbivory is one of the most basic and important relationships in nature. Herbivorous insects alone represent almost a third of all living species (Wehling and Thompson 1999). In natural ecosystems roughly 10% of all plant resources are lost to herbivory (Schoonhoven *et al.* 1998). If one looks around the forest, it is difficult to find a leaf that has not been effected in some way by an animal. Herbivory is an ancient relationship, and its long history has important evolutionary implications. It is sometimes likened to an age old, continuous war. It is often suggested that this 'arms race' between the animals eating plants, and the plants trying to avoid being eaten by animals, may have contributed in a large part to the incredibly high diversity of flowering plants and insects that we see today. The war of herbivory can be seen from either side, the animals and their attempts at digesting plants, or plants and their ways of avoiding getting eaten by animals.

Types of herbivores & strategies

Herbivores can be any sort of animal, from the biggest land mammals on earth (elephants), to some of the smallest insects (termites). Herbivory by larger animals is not as important or obvious in the rainforest as it is in the adjacent savanna woodlands. In the tropical rainforest it is the constant attack on plants by the smaller insects that really constitute the major battles of herbivory.

Herbivores can be browsers or grazers. Grazers feed on grass, while browsers nibble on a range of plant foods. Grazers are not as obvious or dominant force in the rainforest as they are in the adjacent savanna where there are large areas of grass to graze. Grazers in the rainforest include pademelons (right), although to feed on grass these animals often have to move to the edge of the forest to feed.

Browsers in the tropical rainforest include various species of possums, with the ring-tails especially being mainly foliovores; they are the equivalent to leaf eating monkeys in tropical Asia, and sloths in the Americas. And like their equivalents, these possums need to digest their cellulose heavy foods, and are thus fairly lethargic animals.

Within these groups there are different herbivorous strategies. 'Generalists' feed on a wide range of species, while 'specialists' concentrate on specific types of plants. Generalists can choose from a wider range of food and therefore can adapt to changes in habitat, however, those animals that specialize on particular plants are usually capable of processing nutrients much more efficiently than generalists (Howe and Westley 1990).

Plant defence against herbivory in the Australian tropical rainforest

Plants are at a distinct disadvantage compared to animals, for obviously they cannot run away or attack their predator. However, they have evolved many of their own defence strategies. We usually divide these into two different types of defences; physical and chemical (although, this is largely an artificial distinction and there is considerable overlap and combination of tactics). Physical defences are often external and usually obvious to human eyes. The simple toughness of leaves and their waxy coating can act as an effective barrier against chewing mouthparts, especially of smaller insects or early larval stages (Norris 1991). Even the simple and common pattern of the serrated edge of leaves may slow down an insect herbivore. Spikes, spines and stinging hairs can also act as a deterrents against herbivory.

Such physical defences are usually more effective against vertebrate predators, with smaller insects simply chewing around them. In fact it may offer advantages as a host plant to insect herbivores, as it protects them against predation or accidental ingestion by large herbivores (Norris 1991)! An example of this type of plant in the Australian rainforests is the Stinging bush *Dendrocnide moroides*; it is well known to cause humans intense pain and is recorded as killing a horse with it's poisonous silica hairs (Norris 1991); yet is routinely used as a food plant by the 'White Nymph' butterfly *Mynes geoffroyi* (Valentine 1991).

Chemical and internal composition is the second type of defence. Chemical compounds in the plant that are used for day to day functions are known as primary compounds; secondary compounds are those that have evolved for other uses, such as defence (Harborne 1997). Resins, tannins and silica can render the plant indigestible, essential oils can acts as feeding deterrents, and alkaloids can act as direct toxins (Norris 1991).

Flowers and pollination

Flower and pollination biology

For first time visitors to the Australian tropical rainforest, flowers are often not as obvious as expected. However, a closer look and multiple visits will reveal there is in fact a wide range of colours and forms. This diversity of forms reflects the wide range of different pollination strategies, for the look of a flower reflects it's pollinator; flowers that rely on animals have strong colours and odours, while flowers that spread their pollen by wind are dull as they don't have to attractive, but are usually small and numerous.

In the grasslands and temperate forests, wind pollination is very common. However, throughout the rainforest of the world, pollination by wind is relatively rare (a good thing for hayfever sufferers). There are factors that work against such strategies in this habitat, including the rain washing pollen out of the air, the high humidity making pollen stick together, and the lack of wind in and underneath the canopy (Turner 2001).

It is animal pollination that is the main strategy in tropical rainforests. On a worldwide scale, it has been reported that 98-99% of tropical rainforest plants are pollinated by animals (Turner 2001). Signals such as colour and odour advertise to the animal that there is a reward of sweet liquid called nectar. The design of most flowers is such that when the animal feeds on the nectar, it inadvertently picks up the male reproductive material (the pollen). When the animal then next feeds on another flower on another plant of the same species, they spread this pollen. A range of animals act as pollinators in tropical rainforest. Birds are probably the best known, such as sunbirds (pictured below) and the largest family of birds in Australia, the honeyeaters. Mammals are another group, with bats being by far the most notable. In the insect class, pollination is recorded from 6 different orders, including the thysanoptera, cockroaches, beetles, flies, butterflies and moths, and the hymenoptera (wasps, bees and ants). Of all these groups, the most important pollinators in tropical rainforest are probably the bees (Turner 2001). Larger bees, such as the *Xylocopa* (pictured below) work the canopy, while smaller bees, such as the tiny fly like trigonids often work the understorey (Turner 2001). A flower in the tropical rainforest may be visited by many of these different nectar-feeding animals, however, the visitation rates and length of stay are not necessarily an indicator of who the most successful pollinator for that plant is (Adam 1997).

Pollination aimed at different animals

The flowers of different species of plants have been naturally selected to entice their most successful polllinators ('vectors') with specific colours or smells attractive to that type of animal. This suite of features is known as the 'pollination syndrome', and includes size, odour, structure (such as length of floral tubes) and of course colour. The latter feature is quite complex, with colours that we humans cannot see, such as those at the ultraviolet end of the light specturm. Such colours and markings may even direct the animal into the flower like runway lights. In theory, by looking at those features of the flower, one can make predictions about what kind of animal pollinates it.

For example, those flowers that have evolved to attract birds will tend to have bright colours, especially red (such as the Grevillea pictured right), and with combinations of pinks and yellows, reflecting the strong sense of sight and colour recognition in birds. Those flowers attracting mammals will tend to be duller in colour but with a stronger odour, such as the bat pollinated Bumpy Satin-ashes, reflecting the mammal's nocturnal habits and sense of smell. Flowers attracting butterflies and moths are quite different from each other, reflecting their differences. Flowers attracting daytime-active butterflies are generally colourful blues, yellows, reds and pinks, while those designed for nocturnal moths are paler in colour and strong in odour, such as the native Gardenias.

Variations

There are many variations on nectar feeding and pollination. Bees have evolved with flowers to conduct 'buzz pollination', the ability to vibrate the anthers at a particular frequency that releases their pollen (Turner 2001). There are also those species that, rather than 'playing the game' and being satisfied with protein poor nectar, want to steal the much better quality pollen. Many bees will actually cut into the side of a stamen to drill out the pollen.

However, some plants still manage to get the pollen thieves to act as pollinators. In melastomid flowers (left), commonly seen on the side of tracks in Australian rainforest, the purple flowers actually have two sets of stamens. The brighter contrasting yellow stamens in the middle produce a 'cheap' pollen, while the real pollen is in the duller purple stamen on the outside,which they have to climb over (and thus inadvertently collect) to get to the 'fake' stamens.

Fruits and seed dispersal

Plants generally want to disperse their seed far and away from themselves. The distance is the plants 'seed shadow'. The longer the seed shadow, the less chance the parent has to compete with it's own seedlings. As with pollination, plants have evolved various methods ('vectors') to disperse their seed, including animal agents, wind and water. And as in flowers, the most successful dispersal agent used determines the characteristics of the seed and fruit. The lack of colourful and showy flowers in the tropical rainforest of Australia is somewhat made up for by the spectacular fruit. There is a huge range of different colours and forms of fruit that can be found lying around on the rainforest floor.

Animal dispersal of seeds

Studies of subtropical and tropical rainforest around the world have shown that fleshy fruits are produced by 70% to 98% of the plant species (Wilson *et al.* 1989, Loiselle and Blake 1999). In the rainforests of Queensland, the percentage of plants with fleshy fruits has been recorded at about 85% (Wilson *et al.* 1989).

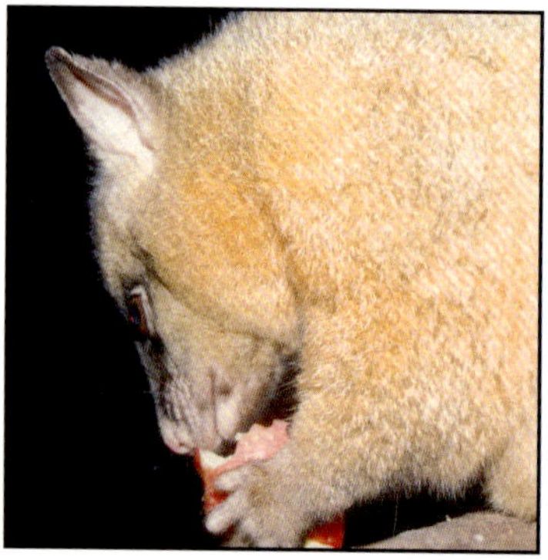

This diversity of fleshy fruits is mirrored with the diversity of fruit eaters. For while there are no specialized fruit eating birds in the whole of the northern temperate region of the Earth, the tropical rainforests of Asia, the Americas and Australasia are rich in such species (Snow 1981). Animals whose diet is made up mainly of fruit are known as 'frugivores'. Animal dispersed fruits advertise, just as flowers do, with various cues such as colours and smells. They reward the disperser with a fleshy edible package of water, carbohydrates and/or fats.

For the plant, using resources to make a fleshy layer around their seed is a sound investment. By using animals to eat their fruit, then swallow the seed and then later void it, the seed is not only dispersed away from the parent tree, but is also provided, by way of the faeces, a microsite of moisture and extra nutrients (Grice 1996). This latter advantage may be very important in the competitive rainforest, especially in those that grow on relatively nutrient poor soils.

However, fruit eaters can vary in their effectiveness as seed dispersers. There has been studies in Australia that measure their effectiveness by the time they spend in the fruiting tree and how far they disperse the seed (e.g.; Green 1993). There are frugivores that are not very effective at seed dispersal at all, such as 'seed thieves' who eat the fruit, but then drop the seed at the site of the providing parent tree, therefore not effectively dispersing the seed.

And then there are those animals that feed on the seeds, the 'seed predators'. In Australian rainforests, the parrots are probably the best known seed predators, ripping through woody fruits with their sharp beaks to consume seeds, such as *Flindersia*, which, despite their hard shell and sharp 'tubercles' can often be seen in pieces on the ground, chewed up by cockatoos. However, one type of 'seed predator' may manage to disperse the seeds over time. There are animals that bury many of their seeds for later use. They are known as 'scatter-hoarders' and their dispersal works through the idea that they may forget to retrieve the occasional buried seed. Here, it may be protected and safe from some other seed predators and thus effectively dispersed.

Examples of scatter-hoarders in the Australian tropical rainforest include the White-tailed Rat and the Musky Rat Kangaroo. However, studies indicate that they retrieve virtually all of their caches, and thus seed mortality is virtually 100% (Strahan 1998).

Different types of fruits aiming for different types of fruit eaters

Fruits are dispersed by an array of animals, including birds, mammals, and even fish and reptiles. Seed dispersal by reptiles is called 'saurochory', and in the rainforests of Australia, the eating of fruit by Boyd's Forest Dragon has been recorded (Torr 1997). Some fruits have evolved to be eaten by a variety of animals, while others are aimed at a specific type of disperser. The type of dispersing agent can be determined by examining the features of the fruit. 'Birds fruits' (below left) are usually brighter coloured as birds have an excellent sense of vision. The fruits are usually smaller as birds to need to retain lightness for flight, have no teeth and swallow the entire fruit, and lack the guts to break down larger fruits. They usually have a single seed. 'Mammal fruits' (below right) don't necessarily have a bright colour even when ripe, as most mammals are nocturnal. Most mammals have a strong sense of smell, and so fruits aimed at them often have an odour. Mammals have teeth and chew, therefore there are usually several, smaller seeds, which may pass through by accidental ingestion. However, there is much overlap between these generalized fruit types, especially in a forest where there are large birds that live on the forest floor like mammals, and there are mammals that fly like birds.

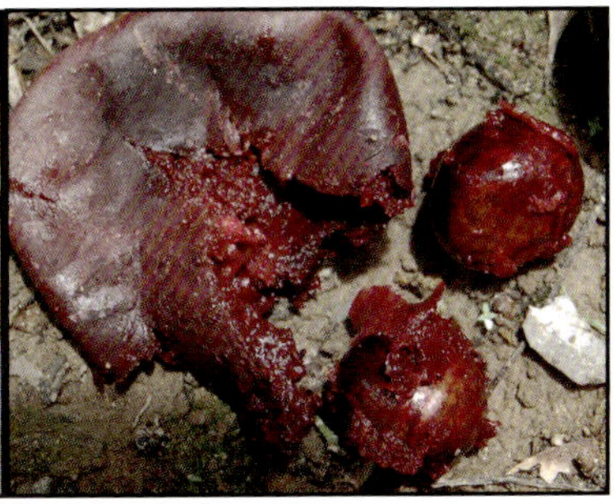

Seed dispersal by mammals

Compared to other tropical rainforests of the world, Australia has a low diversity of mammals. Subsequently, seed dispersal by mammals is a smaller topic. Most noticeably missing from our rainforests, and most obvious in other rainforests around the world, are the monkeys and other primates. The most important group of mammals for seed dispersal in the tropical rainforest of Australia, New Guinea and many of the surrounding tropical Pacific islands are the bats. In some ways, the flying fox are the equivalent in Australia. The traditional division of 'bird' vs 'mammal' fruits becomes complicated when bats are examined, as they have the senses of mammals, but the flight and mobility of birds.

Bat fruits

and it is one of our most interesting and underrated of mammals. Fruits that are attractive for flying fox have certain attributes that appeal to their senses and their morphology. They are often attracted to fruits that are duller or lighter coloured (Richards 1987), juicy and sometimes strong smelling. The fruits themselves are varied in size, and can often be quite large, so large in fact, that most birds can not disperse them (Whittaker and Jones 1994). However, some of the fruits that are successfully transported by bats have many small seeds; as larger seeds will not pass through the smaller oesophageal openings of flying fox, and thus larger seeds will be dropped nearer the tree (Whitaker and Jones 1994). Compared to the oily fruits for obligate frugivorous birds, bat fruits tend to be much more 'sugary' (Whitaker and Jones 1994).

Flying-fox have been suggested by many researchers and their studies to be very effective seed dispersers. This is despite the fact that, at first glance, they would seem inefficient, for flying fox do not necessarily eat a whole fruit. They are often observed to chew on the fruit and will suck out the juice to spit out the pulp and thus often the seeds. However, many of the fruits favoured by such animals (and other mammals such as ourselves) have many smaller seeds that are difficult to separate from the surrounding sloppy pulp. A study on rainforest on islands in the tropical Pacific showed a simple but telling correlation; where there were less flying-fox, there was less seed dispersal (Fuentes 2000). Of course the most obvious advantage to dispersal by flying fox is the fact they can, as the name suggests, fly. They are often noted to roost a long way from their feeding areas, with some species recorded travelling up to 40 kilometres from their roosting to their feeding areas (Kalko *et al.* 1996). The main species of flying fox found in the Australian tropical rainforest (the Spectacled Flying Fox) has been shown to disperse viable seeds for over 26 plants (Hall and Richards 2000), with 23 of these being in the tropical rainforest (Richards 1987).

Seed dispersal by birds

Birds are the best known and the most commonly attracted dispersal agent in the rain forests of the world. One study estimated at least 68% of tree species produced bird dispersed fruit (Wilson *et al.* 1989). And here in the rainforests of tropical Australia, over 40 species of birds have been recording regularly including fruit in their diet (Jones and Crome 1990), including bowerbirds, catbirds, birds-of-paradise, figbirds, orioles, starlings, and honeyeaters (Green 1993). The most diverse and abundant group of avian frugivores are the pigeons; there are 7 species of obligate frugivorous pigeons and several more species that feed opportunistically (Crome 1975).

Plants, fruits and the physiology of birds have co-evolved to some extent. Once the bird has eaten the fruit, they void the seed quickly relative to other animals, as they need to lighten the load. Depending on the species, seeds can take anywhere from 10 minutes to 2 hours to pass through (Green 1993). This would seem to be a disadvantage for a plant trying to disperse its seed, however, birds are fast, active and mobile, and studies have observed that birds generally don't stay for too long in the fruiting tree, flying to another before voiding the seed and thus dispersing the seed effectively (Green 1993).

Fruit-eating birds can be further divided into those specialist fruit eaters ('obligate frugivores') and the more generalist bird that includes fruit as well as other food in its diet. The obligate frugivores that rely on fruit and little else for their diet also tend to feed on fruit that is more nutritious and rich in fats and proteins, such as plants in the family Lauraceae (Crome 1975). The less specialized fruit eaters feed on the less nutritious fruit that are high in sugars and water content (Snow 1981). Obligate frugivores tend to have gentler gizzards, and pass the seeds less altered and thus more viable; this is probably also a way of avoiding the poisonous compounds that are present in many seeds (Stocker and Irvine 1983), which would be released if broken down.

The presence of many species of giant single-seeded brightly coloured fruits suggests there must be a giant fruit eating bird, and indeed the largest native animal of this rainforest is a giant frugivorous bird. The cassowary is probably the single most important animal seed disperser of the local rainforest. So far, over 200 rainforest plants have been recorded to be eaten by cassowaries (Kroon and Westcott 2001). One study alone recorded 78 species found in cassowary dung, with at least some of the seeds of 70 of those species remaining viable (Stocker and Irvine 1983). As with many fruit specialists, many of the seeds come out virtually untouched and many of these are far too large for any other extant animal to disperse (Stocker and Irvine 1983). Cassowaries are effective dispersers over some distance with studies showing territories of up to several kilometres (Kroon and Westcott 2001) and seeds may take over 10 hours to pass through the digestive system, allowing time to cover these distances (Stocker and Irvine 1983).

Wind & water dispersal of seeds

Despite the predominance of fleshy fruits in the rainforest, the diversity of woody fruits in the rainforest is still impressive. It is not uncommon to see hard, often large, woody fruits ('follicles') on the floor of the tropical rainforest, although this is also partly due to the fact that such fruits tend to last longer on the forest floor and are not a food source as are the fleshy fruits. Woody fruits may indicate either wind or water dispersed seeds.

Qualities of wind dispersed fruits and seeds

Wind dispersed seeds tend to be small, light and numerous. They are usually contained in a protective woody follicle that protects the seeds from predation. These woody fruits open to release the many small seeds, many of which may be winged in some way to assist in dispersal. However, in contrast to the often large woody fruits, the thin papery seeds themselves are much harder to find. They are usually small and delicate, and are eaten by seed predators. It is quite common to find the woody fruits on the ground before they have opened, often chewed apart by seed predators such as Cockatoos.

Strategies of wind-dispersed seeds

One of the more obvious features to human visitors in the rainforest is the lack of wind. Thus, wind dispersal would not seem to be a successful strategy. That is, unless the main wind dispersed seeds were only found on the edge of the rainforest as are 'pioneer species', (such as *Commersonia*), or sticking out above the rainforest as 'emergent species', (such as many of the 'Silk Oaks'). As rainforest emergents with wind dispersed seeds, it's a simple equation; studies showed that generally, the taller tree, the further the seed fell away from the parent, or the longer the seed shadow (Nadolny 1999).

In the forests of south-east Asia 1 family dominates the rainforest, both in height and in species; the Dipterocarpaceae. These huge timber trees have winged, wind-dispersed seeds (Whitmore 1999). The rainforests of Australia, although with a very different evolutionary origin and history, have their parallel in the family Proteaceae. While there are many smaller species, sometimes with fleshy fruits, there are many Proteaceae that are also rainforest giants with winged, wind-dispersed seeds, such as the 'Silky Oaks' of various general. Resembling the seeds of the Dipterocarps even more are the seeds of the Red Tulip Oaks *Argydendron* of east coast Australian rainforests. Like many of the Dipterocarps, the large seed is conspicuously winged (pictured below). To avoid being eaten, many of these wind-dispersed plants tend to invest all the resources into one big explosive seeding event, and this peaking is known as 'mast seeding'. This occurs in Australian tropical rainforest with species such as this *Argydendron* (Nadolny 1999).

Water dispersal of seeds

In ecosystems based around water, such as mangroves and other seashore communities, the dispersal of seeds with water is dominant. Thus, the fruits tend to be woody and buoyant. In the rain forest, there are species that produce similarly woody buoyant fruits, and thus by implication are water dispersed.

This is not unreasonable to expect, as rain forests grow in, and are the result of, high rainfall areas, and there are many waterways. Further, there is often a graduated transition in species composition, and thus overlap, between the mangrove and the coastal lowland rainforest. Fruits such as the large seeded pod of the 'Black bean' *Castanospermum australe* (pictured left) float well in water, and in fact are even used by children for doll's canoes (Scarth-Johnson 2000).

Many fruits that seem obviously aimed at other vectors may still nevertheless be capable of water dispersion. For example, there is often conjecture among naturalists that fruits such as the colourful and fleshy *Syzgium* are frequently water dispersed. Their colours and flesh are obvious signals for animal dispersion, yet they are also light and buoyant and can be frequently observed floating along in streams. And in fact, some studies have shown that some fruit that would be assumed to be animal dispersed is indeed often more frequently dispersed by water.

A possible example is the 'Black Palm' *Normanbya normanbyi*, a plant restricted to the Daintree lowland rainforest which produces a large bright red fruit with edible and juicy flesh (pictured right). A study analysed the fruit pattern distribution away from the parent trees, and found they followed lines downslope, suggesting they had been carried by overland water-flow, and most of the fruits in the study had been transported in this way (Lott *et al.* 1995).

Further fruit dispersers once in the water

As these fruits float down along a rainforest stream, they may yet be transported by an animal. In some rainforests of the world, especially those based around the great waterways of South America, some of the over 2000 species of freshwater fish are dispersers of many rain forest fruits (Whitmore 1999). In comparison, Australia has a less diverse freshwater fish fauna, with only about 300 species (Allen *et al.* 2002); nevertheless, fish in local rainforest streams may act as dispersers of seed. For example the Jungle Perch *Kuhlia rupestris* has been reported to eat fruits, such as figs (Allen *et al.* 2002).

Predation

Because of its history of isolation the tropical rainforest (and indeed other habitats) of Australia are notable for lacking the larger placental mammalian predators that are found on the other continents, such as big cats, canids, civets and mongoose, racoons, weasels and bears. The marsupial equivalent to the Carnivora are the Dasyurids.

The largest in this group are the quolls, once known as the 'native cats'. The Tiger or Spotted Tail Quoll is the largest of all quolls, (in fact it is the largest native mammalian predator on the mainland) is a very rare sight in the rainforest, and indeed a very rare sight everywhere except in Tasmania. However, these are relatively small animals, only the size of a domestic cat. It wasn't always the way, with fossil evidence showing that the prehistoric tropical rainforests of Australia and New Guinea once had the marsupial equivalents of leopards and wolves. Today, there are 3 introduced placental carnivores, but none of them are common deep in the rainforest. Two of them, the cat and the fox, tend to patrol the edge of the rainforest, while the third, the dingo and the domestic dog, penetrate further in. This species is now Australia's largest land-based carnivore, and can kill rainforest wallabies and the biggest native animal in the forest, the cassowary.

However, it is important to remember that even in the worlds rainforests where there are native mammalian carnivores, they are in the minority, with a range of other animals eating other vertebrate flesh. In the tropical rainforest (and other habitats) of Australia, some of the reptiles have evolved into the top predators. In the lower reaches of larger rivers, such as the Daintree, the crocodile is the by far the largest predator.

On land in the rainforest proper, snakes are major predators of everything from frogs, to small lizards, birds and mammals. In fact, the Amethystine python is probably the top native predator. Goannas are scavengers of the forest floor, but they also predate on other smaller animals, including climbing trees to steal eggs.

Birds are also important predators in the tropical rainforest. Some species, such as the Black Kite are wide ranging and found in various ecosystems, including on the edge of the rainforest. There are other birds of prey that are more specialized to forest habitats, such as the Crested Baza, and the Grey Goshawk. These birds have shorter broader wings for manoeuvring around and under the rainforest canopy. There are also several species of forest inhabiting owls, including the Rufous and the Lesser Sooty Owl. Some of these bird species have been recorded taking possums from the canopy.

Parasitism

All animals are affected by parasites, and the creatures in the rainforest are certainly no exception. In fact, because they live in moist, humid conditions conducive to life in general, they may catch more than their share. Parasites vary from platyhelminth flatworms, to leeches, insects, and ticks. The latter are commonly seen on mammals of the forest, such as bandicoots and pademelons. And of course we humans are attractive to a range of parasites when we venture into the rainforest, from blood sucking leeches to mosquitoes. Frogs are also an easy group to observe parasites on, as their skin is smooth and naked, and they are out and about on warm, wet nights when parasites are at their peak.

As can be seen by the photograph, frogs are also attacked by mosquitoes, but they also get a range of other parasites. An example of a large parasite on local frogs is provided by the endemic Australian genus of *Batrachomyia* flies. The adult lays an egg under the skin. This then hatches into a larvae that is relatively large compared to the frog host, thus resulting in a hideous looking lump on the back or shoulder. The larvae poke a breathing hole through the amphibian's skin. It eventually emerges, often killing the frog host in the process.

Decomposers

Eventually all organisms in the forest die, and no matter how far up the food chain they are, they too will get eaten. The conditions of moisture and warmth encourage the proliferation of decomposers, and turn over is quite fast. The processes of breakdown ensure dead material is transferred back into living materials. Because of the high rainfall, there is the danger of nutrients, (which in some sandy soil coastal rainforests may be quite low), getting washed away. However, the high rainfall itself contributes to the proliferation of decomposers, so their abundance helps to rapidly break down matter to integrate back into the ecosystem again before it can escape (Osbrone 2000). The agents of breakdown in the tropical rainforest in Australia include: decomposing by bacteria, eating of wood by termites, ants gathering flesh for their larvae (pictured below), brush turkeys picking away at carcasses, and decay by fungi. By far the most important breakdown agent is the latter, with fungi being estimated to be responsible for the release of 85% of the carbon (Martin 2002). There are also associations between mycorrhizal fungi and the roots of rainforest trees that enable the plant to use nutrients in the soil it would otherwise be unable to access. It is now believed that up to 90% of plants in Australia have this type of fungi in their root systems; many of those plants, such as some orchids, cannot survive without it (Martin 2002).

Ecological concepts

Keystone species

Every single species interacts with several other species in the community. Any one species in an ecosystem can be traced to another species through a chain of interactions. But there are certain plants and animals that interact with so many species and in such a way that those other organisms rely on that species for their continued existence, either directly or indirectly. Thus, if they are taken out of the system, they will have a huge effect of many others within that ecosystem. The analogy is with the 'keystone' in a classical arch; if you remove that one brick that is holding everything else in place, it will collapse. A good example of a keystone species in Australasian tropical rainforests is the cassowary.

This animal eats mainly fruit, and in doing so, spreads the seeds of those plants. For many of these large seeded fruits there appears to be few other animals capable of dispersing them. Because the cassowary is a long lived species, when it goes extinct in a local area (as it did around Mt. Whitfield in Cairns over the last two decades), the slow local extinction of plants in the area that relied on the bird may not be noticed for many generations. However, the good news it that to some extent it appears that the more complex a system is, the more robust it is.

Diversity

The diversity in tropical rainforest is well known and recorded. In 1 hectare plots (100 metres by 100 metres), scientists find an average of between 60 to 150 species of trees in lowland tropical rainforest. In very rich rainforest regions, such as western South America, or South-east Asia, they might find over 200 or even over 300 species of tree per hectare (Richards 1996), and in Australian tropical rainforests, over 100 species. Surveys and counts have found a strong pattern of increasing species diversity the further one goes in the tropics. For larger animals such as mammal and birds, there is roughly twice as many species in the tropics as in the temperate regions (Whitmore 1998), and with the smaller animals, such as butterflies, it is even more. For example, in a single tree in Peru, scientists found 43 ant species; more ant species than live in all of Canada (Whitmore 1998), and this pattern is generally reflected in Australia's wet tropics. But why is there so much in the tropics? Suggestions have included the physical structural complexity of the environment and the plants, narrow niches and very specific requirements, climate changes resulting in isolated populations and speciation, and general diversity encouraging diversity. But maybe we are asking the wrong question. Perhaps it is the norm for life to be diverse, and that in fact, we should be asking why there is such a low diversity of species in the temperate and colder areas of the world. That is a much easier question to answer; relatively few animals have evolved to deal with winter and aridity; those that do are adaptable and robust species, and tend to survive over large geographical areas.

Fungi,
Moss
and Lichen

Kingdom Fungi

Fungi actually belong to a completely different kingdom from the animals and the plants and are not particularly related to either. There are thought to be about 250,000 species of fungi in Australia, and thus they far outnumber the 'regular' (vascular) plants.

However, most of these species are too small to be noticed. Only about 5% of fungi produce fruiting bodies that are large enough to be seen by humans. These 'mushrooms' and 'toadstools' (higher image) are only the reproductive parts, which release the dust like spore. The actual main body of the organism is the 'mycelium', which is made up of the hair-like hyphae (lower image) that invade whatever the fungi is living on or in. While the form and pace of life of fungi are obviously plant-like, they do not photosynthesize like plants, but actually ingest food as animals do, either breaking down matter such as soil or wood, or parastising other organisms. This process is accomplished by the creeping hyphae.

By breaking down other matter, fungi play an important role in the recycling within most terrestrial ecosystems, and the Australian tropical rainforest is certainly no different. The fungi and their process is also utilised by the regular higher plants that have mycorrhizal fungi in their root system to extract nutrients from the soil. And in fact, the more it is studied, the more it has been found that other plants have, and rely on, symbiotic fungal associations.

Many fungi remain unidentified, so it is recommended not to consume any wild fungus, even if you think you recognize the species. Identification and taxonomy of fungi is a fairly obscure and specialist task, and many groups are thankfully named obviously and simply for their shape and design of these reproductive structures. With such a diversity of species in the rainforest, this next section can only look at the more commonly seen groups.

Division Ascomycota

This division includes a wide variety of fungus types, including the insect parasitising 'vegetable caterpillars' of the genus *Cordyceps* spp. The group also includes more useful fungi such as yeast and truffle. Perhaps the most obvious group found in the tropical rainforest are the cup fungi. As this common name suggests, these species develop a soft, cup-shaped fruiting body. In the tropical rainforest they include a range of pretty colours, but the most commonly seen are the orange/red cups of *Cookeina* spp., found growing on fallen twigs on the forest floor (pictured right).

Division Basidomycota

This huge group includes most of the fungi that produce familiar fruiting bodies, from the classic toadstools to coral, puffballs and jelly.

(Jelly fungus)

The jelly fungus are often considered a more primitive group (Young 2005). As the common name suggests, the fruiting bodies of these fungi are usually soft and wobbly like jelly. They often grow hanging from standing trees or on logs, or even parasitize other fungi (Fuhrer 2005). They have a high water content and many are known to be edible.

Order Tremellales

The most commonly seen jelly fungi are probably the ***Tremella*** spp. (pictured below).

(True Gilled Mushrooms)

Order Agaricales

This large order includes many species of what most people would call the classic 'mushrooms' and 'toadstools', including the standard edible mushroom. The fruiting bodies have a stalk and cap, and on the underside of the cap there are feathery soft gills, from where the spore is released.

The diversity in colour and form is enormous, for they can range from delicate pin size to plate size, and occur in every colour. Two main groupings can be recognized based on their gills; one group has relatively simple gills, while the other group has forked gills. An example in the first group is the genus *Mycena*, which is made up many dainty, usually white mushrooms, including the bioluminescent fungi of the tropical rainforest, *Mycena chlorophanos*.

__Cryptotrama aspratum__ (pictured right). Another example in the first group of simple gill mushrooms. These small bright red to yellow mushrooms usually occur singly or in small groups. On top of the cap are many tiny distinctive projections with a slight curl, which fall off as the cap expands (Fuhrer 2005). The spore is white. They are usually found on decaying wood in wetter forests of eastern Australia, as well as parts of North and South America (Young 2005).

Marasmiellus spp.

The fungi in this genus produce many tiny mushrooms. They have relatively large, fine but jutting gills underneath delicate caps. Most of the species are mainly white, sometimes light pink or brown, and some of the species smell (Fuhrer 2005). *__Marasmieluus candidus__* (pictured left) produces delicate white mushrooms. They have a fine single stem, and the caps often turn upwards as they age. They usually form on dead wood on the floor of the wetter forests.

Fleshy pore fungi Order Boletales

The boletes are often known as the 'fleshy pore' fungi, as the fertile surface underneath the cap has small holes or pores rather than the gills of the previous group. The fruiting bodies usually have the traditional toadstool shape, and many species are quite large with thick stalks and big caps. They can be quite vivid colours, especially when the flesh is damaged, as the oxygen in the air causes enzyme reactions to make the flesh change colour (Young 2005). Many species are also important mycorrhizal partners for Australian trees (Young 2005).

(Woody shelf fungi)

This group includes many different shapes and forms that are common in the tropical forests, including the 'hard cup fungi' and the 'shelf', 'plate' and 'bracket' fungus commonly found growing off logs and still standing dead trees. The texture of the fruiting bodies of many commonly seen species is hard, leathery or even woody, and they may have a short stem. They often grow in semi-circular shapes from wood, looking like shelving growing out of trees. Their hardness means they are very resilient and can live for years. Many species develop beautiful multi-coloured patterns; the bands or circles of colour are actually annual growth rings, and the whiter outer layer is the growing edge, which sometimes represents a spore layer. Some of the larger species may grow large thick shelves that contribute to the death of the tree, and then feed off the wood for years after. Others may form vital mycorrhizal partnerships with forest trees (Fuhrer 2005).

Order Hymenochaetales

Phellinus spp.

These are very hard and solid shelf fungus; an axe or saw may be needed to split the fruiting body (Young 2005). As they continue to grow over the years, they form new layers. As the shelving grows slightly down over time, it ends up in a hoof shape, or looking like cow pats. There is often a green colour against the dark brown; this is from lichen and moss growth.

Order Poriales

Microporus xanthopus (pictured right) develops medium to quite large 'cups' up to 10 centimetres across. As they grow they develop a series of beautiful coloured zones in rings, from white on the edge, through alternative shades of brown and yellow. They are white underneath, attached by a short base, which is often swollen where it attaches to the wood. The stem and foot usually have some brown, orange or yellow markings, which is acknowledged in the second part of the scientific name '*xanthopus*', which means yellow foot. Their shape and the growth in wetter forests means they often collect water. This fungi is common on small logs and branches on the forest floor of tropical and eastern Australia.

Gasteromycetes

This interesting group contains some of the most bizarre looking of the fungi, from the puffballs to the stinkhorns.

Stinkhorn fungi are often first detected by their bad stench. If that wasn't bad enough, species in the genus *Phallus* are usually very penis-shaped. Some species, such as the Bridal veil fungus, have a delicate lace like network (pictured left), while other Phalloids are star-shaped. The tissue that contains that spore breakdowns into a slime called 'gleba' that produces a rotting smell (Fuhrer 2005). When insects such as blowflies are attracted to the fungus by the 'rotting' smell and consume the slime, they end up distributing the spore on their feet or through their system to other locations. The slime itself is toxic to humans, and it has been reported that dogs attracted to the smell have been poisoned from consuming stinkhorn fungi (Young 2005).

Lichen

Lichen is actually a life-form that is the result of an inter-dependent relationship between two very different organisms from completely different kingdoms. This symbiotic partnership is made up of a 'mycobiont' and a 'photobiont'. The first partner is a fungus, while the second partner is either a green algae or a cyanobacteria (the latter partnership is a notably more common occurrence in the lichens lower down in the rainforest strata, Purvis 2000).

Uniquely, the resulting life form known as a lichen doesn't resemble either of the separate partners. While biologists are still not exactly sure what the true dynamics of the relationship are, generally the algae provides the food through it's photosynthesis, while the fungus provides protection from the elements, as well as mineral nutrients.

In some older classifications, the lichens are placed in the plant kingdom, however, they are really identified based on the fungal part of the relationship, for while many different algae are involved in different lichens, each lichen species has it's own specific fungus (Purvis 2000). The type of lichen-forming fungi can vary, but more belong to the 'cup fungi' of the Ascomycota than any other group (Purvis 1999).

The result of this partnership is a very successful and diverse group of organisms that can live in habitats all over the world, including some of the harshest environments on the planet. There are over 13,500 species so far classified (Kantvilas and Jarman 1999), and many more to be discovered.

Lichens are also abundant in the Australian tropical rainforests, and can be found in the same forms as elsewhere in the world. These sizes and shapes range from the micro lichens that look like paint on the tree trunks, to the splashes on every leaf (pictured bottom of opposite page), to the macro-lichens such as the stringy 'Old Man's Beards' which hang down from tree branches.

In rainforests, lichens contribute greatly to the species diversity. In the tropical rainforest of Central America, single leaves have been found supporting over 50 types of lichen (Purvis 2000). And in another survey, biologists found over 170 species of lichen on one single individual Quandong (*Eleocarpus*) tree in the New Guinea tropical rainforest.

Different types of lichen inhabit different zones within the forest; this can be observed at both small and large scales. On a bigger scale, as the altitude increases and the temperature drops, lichens become even more obvious. In the cooler highland forests of Bartle Frere (or in southern Australia, such as Tasmania), elaborate lichens hang everywhere off the surrounding trees. On a smaller scale, zonation can easily be observed by examining a typical trunk with its different colours of lichen at different levels (pictured right).

Lichens have traditionally been described in various forms from 'crustose' (such as the 'splashes of paint' on trunks), to 'foliose' for the more leafy lichens. However, these descriptions don't correlate with taxonomic categories, so they are not elaborated upon here.

Old Man's Beard
Usnea spp.

One of the most commonly seen and recognizable groups of macro-lichens are the 'Old Man's Beards'. There are over 500 species in this genus and they are found all around the world (Kantvilas and Jarman 1999). Many species grow as bunches of stringy material hanging down off branches, and local examples can be seen hanging from trees over the water at Lake Barrine (pictured left).

Kingdom Plantae

Mosses and allies

Bryophytes

The mosses, hornworts and liverworts are now classified in 3 different divisions, reflecting their likely separate lineage; however, they are often commonly lumped together as the 'bryophytes'. They are generally considered the simplest of those organisms that are actually included in the true Plant Kingdom. They are mainly only one or two cells thick in structure, and most have no efficient system of controlling or moving water within or without their body, in contrast to all the other higher vascular plants. This, and their reproduction, means bryophytes are more prolific in very moist places, such as rainforests, where they grow on the ground, on rocks and on other plants, especially in gullies and around creeks. There are somewhere between 16,000 to 25,000 species worldwide (Jarman and Fuhrer 1995). Of these, there are some 950 species so far recorded in Australia, with about 400 of them found in North Queensland (Jackes and Cairns 2001). As with the lichens, mosses appear to be even more dominant in the cooler higher altitude rainforests, such as those of the Atherton Tablelands.

The bryophytes play important ecological roles in the tropical rainforest. They act as 'sponges' in the forest, catching and absorbing the huge amounts of rainfall, slowing the general flow of water through the system, and thus reducing the rates of evaporation, run off and general erosion.

Mosses and other small sessile organisms such as lichens and fungi are pioneering colonisers of initially bare surfaces such as rock and dead logs, and thus allow for the latter establishment of epiphytes and other seedlings. They physically allow shelter and protection for many of the high diversity of small invertebrates of the forest. They probably also form part of the diet for many of these animals (Jarman and Fuhrer 1995).

From a distance, bryophytes appear only as a blanket of soft green carpet. Close up, tiny individual plants can be distinguished. Generally, they all have small green 'leaves', (although these are structurally not 'true' leaves). The main leafy part of the plant (which is in fact only the half chromosome gametophyte) develops the sporophyte, usually as a thin projection above the plant (pictured at bottom of opposite page), or a small bucket. These structures release the spores. Such closer looks at these plant can reveal relative differences that can be used to distinguish between the main groups.

Division Marchantiophyta

The liverworts are either 'leafy' or 'thallose'. The 'leafy liverworts' tend to look similar to moss with a series of small leaves, but they are arranged in one flatter plane, rather than spirally as in moss.

The 'Thallose liverworts' tend to have single larger flatter leaves at the base, which resemble seaweed and are called the 'thallus' (Jarman and Fuhrer 1995). They have a cup-shaped capsule at the end of their sporophyte.

Division Anthocerophyta

Hornworts also have a seaweed-like thallus at the base (pictured left). Their common name comes from their often double pointed horn-like sporophyte, although this projection collapses much quicker than those growing off moss.

Division Bryophyta

Mosses tend to have a series of small spirally arranged leaves, which are usually a little pointed (right) and often more upright than the previous bryophytes. Also, their matchstick like sporophyte is obvious and erect (as pictured at bottom of page opposite), and usually lasts longer than in the previous groups.

Vascular Plants
Division Tracheophyta

Defining plants

The boundaries between what is and what is not defined as a plant is somewhat blurry. Fungi were once considered plants, but they have been placed in their own kingdom for some time now. Algae are often included with the plants, but more recent taxonomy puts them in various other groups (Ingrouille and Eddie 2006). Lichen are a combination of fungi and algae, but the fungi is what determines their species name, so they are also not considered a member of the plant kingdom. Thus, of the previous entries it is only the bryophytes that are included in the Plant Kingdom; however, moss and their alike are mostly restricted in their growth to just a few cells in thickness and lack the ability to grow into elaborate large forms. In contrast, the 'vascular plants' of the next section include a huge range of forms, from vines to shrubs to towering trees, including the tallest of all organisms that have ever lived on Earth.

Importance of plants

Plants represent arguably the most important group of organisms for several basic reasons. They provide the structural basis in terrestrial ecosystems, and this is especially important in rainforests where plants grow on and around other plants, as epiphytes and climbers. Plants also provide the basis of the ecological food pyramid; herbivores feed on plants, and these animals in turn provide food for animals that eat them, thus all other organisms ultimately rely on plants. Plants also comprise the underlying and basic species diversity for ecosystems around the world, and again this is most obvious in tropical rainforests; this is reflected here, for they make up by far the biggest section in the book.

Diversity of vascular plants

Tropical rainforests, especially those in the lowlands, are usually recognized as the most species rich of terrestrial ecosystems. This is particularly so in the area of plants (and insects). And in Australia, the general trend of high diversity in rainforests is also apparent. The highest diversities on land for most organisms per area, including plants, are in the north-east, with the only serious competition to this diversity of plants being the heathland communities of south-western Australia. To the causal visitor there is no distinction between all the different species; the rainforests appear as a massive wall of impressive green from the outside and a huge jumble of plants upon plants from the inside. There are so many types that they confuse even the most dedicated of naturalists, let alone the casual observer, and it is rare for even professional local botanists to be able to instantly identify all the plants in a given patch of rainforest. Because of their shared structure, rainforests all around the world look quite similar, but the more one gets to recognize species or even groups, the more their differences become apparent.

Plants

(Fern-allies)

'Ferns allies' is a loose term to describe a collection of plants that are not all necessarily related to each other, but united in being the most ancient of the vascular plants. The following groups of plants have a long fossil record, and back in the Carboniferous period, even before the dinosaurs, there were species that grew as large as trees and comprised the dominant vegetation. Despite their ancient giant size and their evolutionary importance, today they exist only as modest and rarely noticed species. They are however found all over the world, from the arctic to the tropics, and are well represented in the Australian tropical rainforest. They differ from ferns by lacking large leaf-like fronds, and by the fact that the spore cases are found on the terminal ends of the leaves rather than on the underside, as in most fern species (Jackes and Cairns 2001).

Sub-division Lycophytina

Class Lycopsida

Tassel Ferns and Club Moss
Huperzia/Lycopodium spp.

Despite the common names, 'Club Moss' and 'Tassel Ferns' are not related to moss or ferns, although they resemble the latter. Their tiny leaves, called microphylls, are almost scale like. 'Tassel Ferns' hang down as epiphytes from trees in the canopy (below, right), while other species of 'Club Moss' grow from the ground (below, left), like small ferns. At the end of the stems are the collections of spore containers; in the tassel ferns they can be quite long and hang down (below right), and in the club moss they stick up like little paws (below left), and thus the name *Lycopodium* means 'wolf's feet'. They are well represented in the rainforest of tropical Australia, where there are at least a dozen species, including 3 species that are endemic to North Queensland (Bostock and Field 2007). Tassel ferns can be seen growing off trees along the Mossman River, while the ground dwelling club moss can be seen along the hillslopes of the Bloomfield track.

Class Selaginellopsida

Electric Ferns
Selaginella spp.

There are some 700 species in this group, the sole genus of the class (Ingrouille and Eddie 2006). They resemble ferns in their habit of growing on the floor of the forest, often in stands, and sometimes as epiphytes, and also in the look of their leaves.

A closer inspection will reveal that there are actually two sets of leaves, with a row of much smaller leaves along the main stem. Some species reflect a bluish-green sheen, although this seems stronger in tropical Asian species; the colour may be due to the use of a slightly different part of the light spectrum in the relatively lower light of the understorey. Their subtle beauty is reflected in the name 'Peacock fern' in tropical Asia (Yeow Chin 1998).

Sub-division Psilotophytina

Class Psilotopsida

Fork Ferns, Whisk Ferns
Psilotum spp., *Tmesipteris* spp.

There are between 3 to 8 species of these fern-allies found around the world (Ingrouille and Eddie 2006). *Psilotum* are small wispy plants that can be found growing as epiphytes or lithophytes in the mangroves, rainforest or monsoon forest of tropical Australia (left). They have very long and thin, light green stems that 'fork' repeatedly (thus the name), and barely noticeable scale-like leaves. Along the stems they have tiny yellow spots; these contain the spores. The stems continue to bury as root-like structures called 'rhizomes' into the trunk of the tree host or between rocks. They can be seen growing from mangroves along Marrdja boardwalk, around Lake Barrine, and in the monsoon forest around Undara lava tubes.

Tmesipteris are quite different looking; they resemble in form and habit the hanging tassel ferns of the previous page. They are most common hanging off tree ferns, cycads and other trees in temperate rainforest gullies, although a few are also found in the tropical rainforests of Australia (Jones and Clemesha 1981).

Ferns
Sub-division Polypodiophytina

Ferns are an ancient group of plants that preceded the evolution of the now dominating flowering plants. They probably reached their peak in importance around the Carboniferous period. The parts of the plant have different names than their functional equivalents in the seed plants and can get very confusing; so in this section less technical and more general names are often used. The leaves of ferns ('fronds') have the distinctive feature of unfurling from a rolled up state known as a 'crozier' (commonly called 'fiddleheads', pictured below). Fern fronds have evolved many functions, including photosynthesis, but in some species, particularly epiphytes, non-green leaves are used for holding the plant or collecting debris, and in all species the leaves are the launching pad for the next generation of ferns.

Rather than fruits and seeds, fern disperse by dust-like spore. These are released from spore cases known as 'sporangia', tiny structures that are usually found on the underside of the frond. They are arranged in groups called 'sori' which usually appear as small rusty coloured circles or horseshoe shapes (pictured right). Each different type of fern has a different pattern of arrangement of the sori, and this assists in species identification.

Unlike the more advanced and dominating seed plants, ferns generally do not develop tall, multi-branched growing forms with large woody trunks, and are usually limited in size, being mainly found in the understorey. Like all the previous groups of relatively primitive plants, ferns usually require a moist environment in which to reproduce. Thus, if there is any habitat where they do well, it is in the rainforest, and in the wettest part of Australia, north-east Queensland, the ferns are a very obvious part of the habitat. The forests here have the highest diversity of ferns on the continent, with over 250 species, and of these about a fifth are found nowhere else (Bostock and Field 2007). They vary from tiny plants that cling to the roadside, to huge epiphytes, to towering tree ferns.

Class Marattiopsida

Family Marratiaceae

While the tallest ferns in the Australian tropical rainforest are the Tree Ferns, the largest fronds belong to the magnificent King Ferns of the genus *Angiopteris*. This genus appears to be very ancient and relatively unchanged, for *Angiopteris* fossils have been found that are several hundred million years old.

King Fern *Angiopteris evecta*. Huge size has resulted in them also being known as 'elephant ferns" in Asia. Enormous size means they have to live in very wet areas to retain turgor to keep their fronds erect; thus this plant is very restricted in where it can live; usually only near waterfalls, along creeks and in gullies in well-developed rainforest along east coast of Australia. Also found in tropical Asia.

Modern Ferns
Class Polypodiopsida

Order Gleicheneales

These short ferns grow from rhizomes that creep along the ground. The fronds continuously grow; when one set of fronds stop, another pair of side branches starts, and there may be a tangle of many dark, dead stems behind the living stems. They are thus often referred to as 'thicket ferns' as many species form dense shrubs. This doesn't allow for the growth of a lot of other plants, but the thickets are not impenetrable, as the stems are quite slim, and the ferns and creeping rhizome can be destroyed by fire (Chin 1998).

These ferns can colonize and take over open areas, so they do well in disturbed areas, such as rock slides, the sides of roads, and the edges of gardens and other urban areas. Most species are found in the tropics, at various altitudes, and genera include *Gleichenia, Sticherus* and *Dicranopteris* (pictured left, along the Cape Tribulation road).

Tree ferns
Order Cyatheales

Amongst the prettiest and spectacular of ferns are the 'tree ferns'. They are true ferns, but the main stem and it's tough roots develop as a single tall trunk.

The latest fronds erupt attractively from the growing end of the trunk; in many species the previous stem attachments can be seen as scars down along the trunk. Being attractive means tree ferns often serve as a symbol for the tropical rainforests around the world, but can in fact they can be just as common, if not even more abundant, in the cooler subtropical and temperate rainforests and woodland gullies of the southern hemisphere. There are 2 families found in Australia; but most species seen in the lowlands and mid altitude rainforests of North Queensland are of the genus *Cyathea*.

Cyathea spp.

With over 600 species, this is the second largest genus of ferns, and they are found all around the world. There are about a dozen species in Australia (Jones and Clemesha 1981).

Rebecca's Tree Fern *Cyathea rebeccae* (pictured below). Very thin tree fern, often not that tall. Dark green fronds tend to stick up at slightly uneven angles. Distinctive habit of retaining the old branches that hang down as thin brown straps messily underneath the current green fronds. Found at range of altitudes in rainforests of north-eastern Queensland and also extending slightly into Flores (Cronin 1989).

Cooper's/Scaly Tree Fern *Cyathea cooperi*. Tall thin tree. Usually at least dozen light green fronds erupt out in pleasing symmetry. Thick trunk (for tree fern) covered in large scales from previous frond bases, noticeable in flash photograph (below, Cairns Botanical Gardens). Naturally grows in gullies and slopes in forest along east coast of Australia, but one of few native tree ferns that can grow well in more open situations, thus also commonly planted around gardens, and seen on edge of rainforest.

Order Schizaeales

Family Schizaeaceae

Lygodium spp.

There are about 40 species in this genus, with 4 found in Australia (Jones and Clemesha 1981). These distinctive small ferns grow on a long, wiry and continuously growing stem that creeps along the ground and climbs trees. The most distinctive feature of this plants are the small projections from the sides of the fertile leaflets (pictured left), under which are the sporangia. These plants usually grow as climbers around mangroves and the edge of the lowland rainforest.

Order Polypodiales

Family Athyriaceae

Callipteris prolifera. Ground fern with distinctive small leaflets growing from base of true larger leaves (larger photo right); when these taken off, can then grow as plants (Cronin 1993), a fact referred to in second part of scientific name. Sporangia pattern is made up of many small lines that cover almost entire back of frond (insert photograph right). Grows only in shady, very wet areas in rainforest of north-eastern Queensland. Genus name means 'beautiful fern'.

Family Adiantaceae

Maidenhair Ferns
Adiantum spp.

This genus contains over 200 species, centred mainly in South America, with at least 10 species found in Australia (Jones and Clemesha 1981). Maidenhair Ferns are usually recognised by their delicate fronds, which are made of many small and delicate rounded edged triangular leaflets. The frilled edge of the small leaflet covers the sori underneath. The stems are usually distinctly black and wiry, almost polished looking. Maidenhairs usually grow in the wetter situations of forests, such as around creeks, gullies and rocks.

Family Asplenianaceae

Asplenium spp.

There are over 700 species in this, the largest genus of ferns (Warren *et al* 1999), and there are more than 25 species found in Australia (Jones and Clemesha 1981). They vary greatly in size, shape and habit, including both terrestrial and epiphytic forms. The genus also includes the attractive maidenhair-like 'Spleenworts'.

Asplenium nidus and/or ***Asplenium australasicum***. Grow on both rocks and other plants. Can grow to enormous sizes in canopy (pictured left). Shaped like bucket to catch falling debris. Leaves are long and strap like, and dead leaves hang down from the base. Can be seen everywhere in rainforest and mangrove canopy; especially common in lowlands.

Family Polypodiaceae

Felt ferns

Pyrrosia spp.

There are about 100 species in this genus, and they are centred mainly in tropical Asia, but extend throughout the tropics of the world, with at least 5 species reaching Australia (Jones and Clemesha 1981). They have different common names depending on how they grow and their shape. The many species that have long single leaves and grow up or hang down from branches or rocks and are usually called 'felt ferns' in Australia. The species with much shorter rounded non-fertile fronds are called 'dragon scales'. On the strap like species, the sori crowd both sides of the midrib on the underside of the frond, turning the tips brown (pictured below). A closer look at the fronds may reveal small scales or hairs, especially around the sori; this reduces the escape of moisture and allows many species to thrive in more open or drier situations (Jones and Clemesha 1981). They are found growing on tree and rocks in rainforest, monsoon forest, woodland gullies and in streets and gardens. There are some similar looking species of *Microsorum* that also grow as small erect straps off rocks, but the sori pattern is less ordered and dense.

Basket Ferns
Drynaria spp.

There are about 20 species in this genus, with at least 3 found in tropical Australia (Jones and Clemesha 1981). Basket ferns or 'Oakleaf Ferns' have similar habits to the Bird's Nest fern, growing as epiphytes high in trees and catching debris falling down.

Like the staghorn ferns, they develop two very different types of fronds that serve different purposes. Their fertile fronds are green and erect and quite slight, while the basal leaves are brown, short and fat, and catch and retain debris. Basket ferns are common in rainforest in tropical Australia and Asia, but they also grow on rocks in gullies in much drier habitats within this range.

Staghorn/ ElkhornFerns
Platycerium spp.

This is a group of about 15 species (Warren *et al* 1999) found in rainforests throughout the tropics, where they grow as epiphytes on trees. There are at least 4 species in north-eastern Australia. They develop with two distinct sets of fronds. The first set are brown and plate like, growing flat against the trunk; they are sterile and hold in debris to support the plant. The other fronds are green and antler-shaped, and serve for photosynthesis and to release the spore. Species of the genus in tropical Australia look similar to each other, with the green leaves growing out and splitting, thus looking like the antlers that develop on male deer, and giving them the common names. Some species are commonly called 'Elkhorns' although there doesn't seem to be any consistency in naming, and therefore there is no biological distinction or significance between the names 'staghorn' and 'elkhorn'. (And in fact, in the case of 2 of the more common tropical Australian species they have actually been named opposite to what they should be; the so called 'elkhorn fern' has bifurcating fronds more like a regular deer stag, while the common staghorn fern has a larger base and 'fingers' much more like an elk antler; thus it is probably best just to call them all 'staghorn ferns'.)

Seed Plants
Sub-division Spermatophytina

This huge group contains all the seed bearing plants, including the conifers, cycads and flowering plants.

Gymnosperms
Class Pinopsida

This class includes the gingkoes, gnetophytes, cyads and conifers. Only the latter two groups are found in the rainforests of Australia.

Sub-class Cycadidae

Cycads were one of the dominant plant forms back in the Mesozoic, the 'Age of the Dinosaurs', when they would have comprised much of the forest as flowering trees do today. In the present day, the less than 200 species that survive in disjunct distributions scattered throughout the tropics and subtropics of the world probably represent relict populations (Jones 1998). Cycads are slow growing plants that usually have a thick trunk with the new leaves coming out only from the top.

They superficially resemble palm trees; however, they are not true flowering plants, but gymnosperms, and reproduce with cone-like structures (pictured at right). Transfer of the pollen is through wind, bees or, most significantly, weevil beetles, the latter of which were probably some of the first ever pollinators on the planet. Of the seed plants, the cycads are the only ones with motile sperm. This and other primitive characteristics places them as the earliest of the still surviving seed plants that now dominate the planet (Jones 1998).

It has been suggested that other aspects of their current biology is a reflection of their ancient lineage; the lethal compound macrozamin is found in most cycads, (including the species found in the rainforests here), and it has been hypothesized that this high toxicity initially evolved in this group to deter predation by dinosaurs. Despite this toxicity, there are reports of indigenous people eating the roots and treated seeds of these plants from around the tropical Pacific, including in the rainforests of Australia (Low 1997). North Queensland has representatives from all 3 families of cycads, and in the tropical rainforests here can be found the smallest of all cycads (Bowenia), growing underneath one of the largest (Lepidozamia).

Family Zamiaceae

This is the largest of the 3 families of cycads.

Lepidozamia spp.

There are only 2 species in this genus and they are both found in eastern Australia (Jones 1998), with the species below found in the tropical rainforest.

Hope's Cycad *Lepidozamia hopei*. Looks quite a bit like a palm and easy to mistake for one. Tallest specimens growing their single straight trunk up to 20 metres. Normally slow growing. Female produces large cone at top of tree that then tilts over and eventually releases red seeds. Found in forests only in North Queensland. Impressive specimens can be easily seen alongside the Marrdja and Jindalba boardwalk in the Daintree.

Family Stangeriaceae

This is a very small family of 2 genera and 4 species restricted to southern Africa and eastern Australia (Jones 1998). The cycads in this family somewhat resemble ferns and in fact the first species in Africa were initially classified as such.

Bowenia Ferns, Byfield Ferns
Bowenia spp.

Despite the common name and appearance, these plants are not really a fern, but a cycad. The leaves are very shiny, sometimes appearing almost plastic, with new leaves often curled. Being a cycad, they are highly toxic, for they contain the compound macrozamin (Jones 1998), and are probably one of the most toxic plants in the Australian rainforest. The genus was named after the first governor of Queensland, George Bowen. There are 3 species in the genus *Bowenia*, and they are all restricted to tropical rainforests (or wetter gullies of woodland) in eastern Australia (Jones 1998).

***Bowenia spectabilis*.** Common plant of understorey tropical rainforests in Australia, especially at lower altitudes. The species is the more widespread and common of the *Bowenia* genus in North Queensland. Another species is found at high altitude in the Tinaroo Hills on Atherton Tablelands; similar except for slightly toothed margins (Jones 1998).

Conifers
Subclass Pinidae

Conifers grow in surprising variety of forms, from shrubs to the tallest organisms on the planet. Leaves vary greatly, from the 'primitive' more regular looking leaves of local Agathis to the needles of the 'advanced' pines of the northern hemisphere forests. The most distinctive features of the conifers are, as the name suggests, their cones, used for pollen release and capture. They then have a range of seed dispersal, from light wind dispersed seeds to fleshy animal-eaten berries. Like the cycads, the conifers reached their highest diversity in the Jurassic, yet they have survived to the present day well, and although they are not diverse, with only a few hundred species worldwide, a small number of certain types dominate many forests of the northern hemisphere. In Australia there is only a limited number of conifers, but the native species are of interest due to their ancient lineage and unusual appearance. Quite a few of these are found in North Queensland, and have evolutionary connections to other Gondwanan conifers found in New Guinea, New Caledonia, New Zealand, and South America.

Order Araucariales

Family Araucariaceae

Kauri Pines
Agathis spp.

Kauri Pines belong to a genus of about 15 species found around the south-west Pacific (Rowell 1996), including New Zealand and North Queensland. In the North Queensland rainforest, they seem most common at the mid-altitudes, and there regular specimens dominate the canopy. They are usually tall trees with massive, straight trunks that don't branch until near the top (pictured below, left). Their long narrow leaves can usually be found scattered around the base of the trunks. There are several species found in the rainforests of North Queensland. These species normally have flaky bark. The flowers and male cones are relatively small and thin (pictured below, right), while the female is a large knotted green cone the size of a small pineapple. The best known individual examples of Kauri Pines are the 'twin' pines located at Lake Barrine. Kauris are considered important timber trees, and there are also examples of the impressive timber at the Lake Barrine teahouse.

Araucaria spp.

The *Araucaria* was named for a tribe, the Araucani, in Chile, South America where the first species in the genus was found (Rowell 1996). They are also found in New Caledonia, where they are at their most diverse and dominate many natural landscapes, as well as the rainforests of tropical New Guinea and Australia. This southern hemisphere distribution is thought to reflect their Gondwanan history. There are at least 15 species, including some of the most distinctive of all trees, such as the Norfolk Island Pine *Araucaria heterophylla,* ubiquitous to so many of Australia's seaside settlements. There are 2 species that naturally occur in Australia, and both can be found in the tropical rainforest.

Bunya Pine *Araucaria bidwillii*. This tree usually develops a recognisable silhouette, especially when grown in the open, of a general pear shape with a wide base (pictured above left). The female produces what is probably the largest of all the conifer cones (Rowell 1996), a large green, vaguely pineapple-looking cone. This provides edible seeds, famous for once attracting feast gatherings of Aboriginal people from hundreds of miles around. Found at higher altitudes in subtropical rainforest and on edge of forest. Occurs naturally in disjunct populations in two major regions; most common in south-eastern Queensland, especially around Bunya Mountains National Park, but also in a few areas in north-eastern Queensland. Also popular street tree, but have to be careful of falling cones; can be seen in Atherton Tablelands.

Hoop Pine *Araucaria cunninghamii* (pictured above right). Tall very straight trunk. Leaves tend to bunch together at end of branches creating distinctive silhouette, look rather like pom-poms from a distance in some situations, e.g.; Atherton tablelands. Found in scattered populations in tropical rainforest, subtropical rainforest and rocky woodlands (such as Magnetic Island), from NSW, along east coast of Australia and up into mountains of New Guinea. Also grown in plantations, often adjacent to rainforest, such as Dunbulla drive around Lake Tinaroo; considered one of Australia's most valuable softwood timbers (Rowell 1996).

Family Podocarpaceae

This comparatively large (for conifers) family has over 170 species. They are found mainly in the southern hemisphere, from tropical to temperate regions. They are dispersed along the east coast of Australia, with some 16 species, including half a dozen species in the tropical rainforest (Cooper and Cooper 2004). They grow as shrubs or trees, and usually produce fleshy fruits or attachments to the seeds.

Sundacarpus spp.

There is only one species in this genus.

Black Pine, ***Sundacarpus amara***. Tall tree, sometimes with buttress roots. Leaves are simple and long. Bright red, round fleshy seeds, often have black bruises, some fruits have pointed ends, single seed is round, dark brown and smooth. Found at high altitude rainforests in Australia, e.g. Hypipamee Crater, and also found in New Guinea and south-east Asia (Cooper and Cooper 2004).

Brown Pines, Yellow Woods
Podocarpus spp.

This genus has over 95 species, found in South America, Africa, Australia, and Pacific islands (Cooper and Cooper 2004). There are 8 species native to Australia, and half of those can be found in the tropical rainforest. Podocarps can grow scale-like or long, thin leaves with parallel venation; the local species have the latter (pictured below).

They usually produce distinct fruit, with a seed hanging down from a fleshy section; the genus name refers to his feature, with *Podocarpus* derived from Greek words for foot and fruit. The fleshy parts are often reddish in colour in the Australian rainforest species. Juveniles pines can be obvious alongside certain tracks, such as on Jindalba boardwalk (pictured right).

Flowering Plants
Class Magnoliopsida

The diversity and evolution of the flowering plants

There are at least a quarter of a million species of flowering plants. They continue to be discovered; during much of the 1980's and 90's several thousand new species were still being described every year (Cranbrook and Edwards 1994). They are the major form of plant type all over the world; indeed they appear as the dominant of any life form in most terrestrial ecosystems. And nowhere is this dominance more obvious than in the tropical rainforest, with the Australian tropical rainforests no exception.

The diversity of flowering plants in the region is reflected in a complex taxonomy, with a confusing array of different species and different ways of representing them in a book like this. In line with the rest of this guide, they are presented in approximate evolutionary order. The order used is based on the latest in genetic based techniques (which can be different from previous systems); the continuing data produced by the Angiosperm Phylogeny Group (APG).

These tropical rainforests are significant in an evolutionary sense. The tropical rainforest of the south-western Pacific region, and the Australian 'Wet Tropics' in particular, are thought by many scientists to have more representatives of the most primitive flowering plants than anywhere else in the world.

Many first time visitors to the rainforest expect to see many large and colourful flowers. This misconception is provided and sustained by the collections of colourful flowers that are usually presented in tropical botanical gardens. However, many of the native plants in the tropical rainforest have very small and quite dull coloured flowers - but this doesn't make them any less interesting.

Flowering plant biology

As suggested by the common name, the plants in the huge division Anthophyta house their sexually reproductive parts within the structures that we call flowers. Flowers have evolved to release and accept pollen. They may use wind or animal to carry the pollen from their flower to another. If they manage to get pollinated, the seed will begin to grow. Most of the flowering plants develop a container around the seed called the fruit. This provides another name for the flowering plants, the 'angiosperms'.

The flower is a sexual organ (remember that next you shove your nose into one and sniff it). The diversity of the flowering plants is reflected in their diverse combinations of sexual arrangements. Individual flowers can have both female and male 'bits', or be of just one sex. When the latter, and the sexes are in different individual flowers, those flowers can be on the same tree at the same time, or on the same tree at different times, or on different individual plants.

The flower itself is made up many different parts with many different names. The parts may vary considerably in shape and look, and it can be confusing recognizing the different parts of a flower. However, these variations in shape and number of features has traditionally been the main way to identify species, and to classify and determine species relationships.

The largest, best known, most obvious and commonly recognized part of the flower is usually the collection of features and structures that make up the 'perianth'. This is made up of the lowest layer called the 'calyx', which is made up of the sepals, and the next layer up, the 'corolla' which is made up of the petals. These are non-sexual parts, and are mostly used as visual aids to attract pollinators. The numbers and look of the sepals vary, but they are mostly just greenish. They may fall off soon after the flower has opened, or they may stay until the flower has developed into a fruit, in which case they form the 'persistent calyx'; a distinctive ring of lobes or scars on the fruit, common in species such as *Syzygium*. The petals of a flower are usually more colourful, and they vary from thin and pen shaped to elaborate and luscious, as in the orchids.

The sexual parts of the plant are also well known, but are bit more confusing. The male parts of the flower are known as the 'androecium' and made up of the usually pencil-shaped stamens (these are usually long and stick out of the flower, so it is easy to remember that they are the boy bits).

There are usually many stamens, and they located just inside the petals. The stamen itself is made up of the stick-like 'filament' and the 'anthers' at the end. The anthers contain the pollen grains. When the pollen mature, the anther will split and be ready to release the pollen. Pollen may be light and wind dispersed, or it may be collected by an animal. The pollen grains produce two male sperm cells (remember that when you next get pollen on your face).

The female part of the plant is called the 'gynoecium' (this means 'female house') and is made up of the usually squatter, slightly pear shaped carpel (think of the traditional venus female shape). This is usually located in the centre of the flower. The ovary is inside the structure, the style is the part sticking up, and the stigma is the end. In this *Hibiscus* (right), the carpel is the main central structure, with many reduced yellow stamens hanging off it.

There are many variations with all the flower parts. For example, the stamens may be in huge numbers and be colourful and large, effectively acting like colourful petals to attract animals for pollination, as in many Myrtaceae and *Barringtonia* (pictured below).

Order Austrobaileyales

This is a very recently erected and tiny order of flowering plants.

Family Austrobaileyaceae

This small family contains just 1 species and is recognized as one of the most primitive groups of flowering plants in the world. The pollen closely resembles pollen from a fossil plant of 120 million years ago (Keto 1992). Its unusual status is reflected in the fact that the plant's name is used for an Australian botanical journal.

Austrobaileya scandens. Vine that produces oblong orange fruit. Found only in the tropical rainforests of North Queensland.

(The Magnolids)

Order Magnoliales

Family Myristicaceae

The family has species spread throughout the tropics. Many of these plants have aromatic leaves. The family is probably best known for the type genus *Myristica*, found in Asia and the Pacific, which provides with us with nutmeg.

Nutmeg
Myristica spp.

'Nutmeg' is thought to be the only plant that provides us with two spices from the one fruit; nutmeg comes from the crushed up seed, and mace from the red flesh around the seed.

Australian Nutmeg or **Native Nutmeg *Myristica insipida* and *Myristica globosa***. Small to large tree. Velvet brown fruits, which open up to reveal black seed covered in distinctive strippings of red flesh. Found in various forest types along tropical portion of east coast of Queensland, into N.T, New Guinea and up into many Pacific islands. In North Queensland, are common in lowlands and especially on islands off coast. Several good examples near toilet block at Mossman Gorge carpark.

Family Annonaceae

This huge family of over 2,500 species is found mainly in the tropics of the world, but also extends into eastern North America. The stems of the trees and vines in this family often have a zigzag arrangement to them, and there are frequently aromatic oil glands present in the leaves and other parts of the plant. In the Australian rainforests, the species are usually fairly small trees, and the fruits are often arranged in distinctive bunches. The main exception in the region is an introduced member of this family: **Pond Apple *Annona glabra***, originally from the tropical Americas, hich has become a pest in some of the lowland swamps of North Queensland. Other cultivated fruits in this family include the Soursop, Custard Apple and Rollinia. The family also includes the aromatic Ylang Ylang tree, which is used in perfumes.

Pseudovaria spp.

There are about 20 species in this genus found from South-east Asia to Australia. The 4 species in Australia are all found in the tropical rainforest (Cooper and Cooper 2004).

Pseudovaria froggattii. Small, thin tree that produces flowers and fruits on branches and trunk. Fruits are orange and occur in a distinctive bundle sometimes quite low down on the trunk (pictured left). Restricted in habitat, being found only in rainforest in lowlands of Daintree region. Occasionally noticed when in fruit along Dubuji boardwalk and at Mossman Gorge.

Xylopia spp.

This is the most widespread genus of the family, with over 170 species found throughout the tropics of the world (Heywood *et al* 2007). There are 2 species found in Australia; one is found in northern Cape York, while the other is also found in north-eastern Queensland. This latter species is **Orange Jacket**, ***Xylopia maccreae***, a tree that is usually noticed when the bright red inside out fruit has fallen to the ground (pictured below, Cape Tribulation). Inside were black seeds covered in a light blue-grey aril.

Family Eupomatiaceae

This is a very small family that is restricted to Australia and New Guinea, and has only one genus, *Eupomatia*. The primitive flowers have a strong scent, but only last for a day. During that time, they are pollinated by weevils that transport the pollen while eating parts of the flower. This genus has only 3 species, with 1 species (***Eupomatia bennettii***) restricted to the subtropical rainforest of northern NSW and southern Queensland, and the 2 species below found in the tropical rainforest of the north-east.

Eupomatia laurina (pictured above) grows as small tree. May have slight zigzag pattern to the branches. Flowers appear only for a day, followed by very distinctive cup-shaped, flat-topped greenish fruits that look like fleshy gum nuts (pictured below). Also referred to as 'native guava' as fruits reported to be nice; I don't agree. Grows in different rainforest types from Victoria to New Guinea. Small examples on Kulki boardwalk at Cape Tribulation, and larger ones on track to Hypipamee Crater. ***Eupomatia barbata*** grows as smaller shrub, with smaller but similar fruit.

Order Laurales

Family Hernandiaceae

This small family of some 60 species is found throughout the tropics of the world. There are frequently aromatic oils found in the wood (Heywood *et al* 2007), sometimes creating quite unpleasant smells that give some species the name 'shitwood' in Australia.

Hernandia spp.

This the largest genus in the family with some 20 species. It's distribution is focused in tropical Asia, but species are found throughout the tropics of the world (Heywood *et al* 2007)

Sea Hearse, ***Hernandia nymphaeifolia***. Grows as a small tree. Leaves are heart-shaped; scientific name means they are similar to water-lilies. White fragrant flowers. Fruits are unusual; dark brown almost black nut, ribbed, enclosed in a roomy cover rather like a 'lampshade' (Cooper and Cooper 2004). Found in forest and other habitats, especially along coast, tropical Australia, throughout tropical Indo-Pacific to Africa.

Laurels, Native 'Walnuts'
Family Lauraceae

This family has at least 50 genera and 3000 species (Hyland et al 1999) and they are found all around the world. The species in this family are very variable in their features. They can be vines or shrubs, but are mostly trees. The leaves are usually simple, often discolorous, with a paler underside due to a waxy coating that can be rubbed off (pictured below). The leaves usually have many small oil dots, (although they can generally only be detected with a magnifying glass) which provide some species with distinct aromas when crushed. The flush of new growth of leaves can be striking whites, pinks and reds, a contrast to the monotonous green of the surrounding rainforest. The flowers of many species in this family are often small and dull in colour. In contrast, however, the following fruits may be large and are sometimes colourful, with thick flesh and a large single seed.

The most important and well known members of this family would be camphor and cinnamon from Asia, bay laurel from Europe, and avocado from the Americas. The leaves of the original 'Bay Laurel' *Laurus nobilis* has been used as symbol by many civilizations, including for victory and honoured poets (and thus the term ' laureate'). The name 'Bachelor' for a study degree (and meaning a single man) came from the Latin for laurel berry (Perry 1977). Many species around the world were and remain important timber trees. The family is probably best represented in the rainforests of the Americas and Asia. In Australian tropical rainforests, there are many species of what we commonly call 'native walnuts'. These have no relation to real walnuts, but are usually large fruits with a single solid seed surrounded by flesh, and a leathery smooth skin, and look more like avocados, of which they are related to. The family Lauraceae is also of interest here, because it is one of the so-called 'primitive flowering plant families' that are so talked about in regards to the Australian rainforests.

Walnuts
Beilschmiedia spp.

This genus is comprised of some 250 species found throughout the tropics of the world, and includes some valuable timber trees (Heywood 2007). There are 10 species in the Australian tropical rainforests (Cooper and Cooper 2004)

Yellow Walnut *Beilschmiedia bancroftii*. Named for it's yellowish, brownish or greenish fruit, usually size of ping-pong ball, with swelling at one end and point at other (larger photograph at right - scattered around grounds at Heritage Lodge, Cape Tribulation). Seed is distinctive, for has two sharply pointed ends connected by a ridge (smaller photograph at right, Cassowary House, Kuranda). Found only in tropical rainforest of North-eastern Queensland.

Walnuts
Cryptocarya spp.

There are some 350 species in this genus, and they are found throughout the tropics of the world (Heywood 2007). There are over 36 found in the Australia rainforest (Cooper and Cooper 2004).

Poison Walnut *Cryptocarya pleurosperma*. Leaf is distinctive, rounded, with 3 main veins. Fruit is commonly seen on ground in undisturbed rainforest; bright red, and remaining seed is distinctly ribbed. As name suggests, very poisonous plant, with even the sap from tree said to be dangerous to timber-getters. Often has hole chewed into it by the unstoppable White-tailed Rat. Common fruit in the Bartle Frere valleys.

Walnuts
Endiandra spp.

There are some 100 species in the genus *Endiandra* are they are found mainly in the tropical rainforest of tropical Asia, the Pacific and Australia (Hyland *et al* 1999). About 38 species are found in Australia, and of these, 31 have been recorded in the tropical rainforest (Cooper and Cooper 2004). The flowers are only usually small, while the 'walnuts' themselves can be quite large fruits. These fruits are often brightly coloured and contain a single large seed.

Hairy Walnut ***Endiandra insignis***. (pictured below, at Daintree Hideaway). Tree. Usually noticed when produces large bright red cricket ball type fruits. Common name comes from tiny hairs on leaves that can give twigs and veins on underside a golden colour (there is no hair on fruit). Fruit varies, sometimes being yellow or orange, but often very bright red, with chips and scars on outside and suggestion of ribbing. Large single seed is distinctly pitted, with swelling at one end, a often a suggestion of a single ridge. Found only in tropical rainforest of north-eastern Queensland.

Noah's Walnut ***Endiandra micronuera***. (pictured left, on Cape Tribulation road). Can grow to large tree. Usually only noticed when drops its large oblong, egg-shaped yellow fruit. May only fruit every few years. Very restricted distribution; found only in undisturbed lowland tropical rainforest within Daintree region. May be seen along the Marrdja and Jindalba boardwalks when in fruit.

Black Walnut, Queensland Walnut *Endiandra palmerstonii* (pictured below, at Hypipamee Crater). Large tree, sometimes with buttress roots. May exude a black sap (Cooper and Cooper 2004). Produces large cricket ball sized fruit, yellow to green, slightly ribbed. Species is named after explorer Christie Palmerston. Found in rainforests at most altitudes, in north-east Queensland south of Kuranda, especially volcanic soils (Hyland *et al* 2002). Important timber species, marketed as 'Queensland Walnut' (Hyland *et al* 2002)

Dodder Laurel
Cassythia spp.

There are about 17 species in this genus, and they are found mainly in the old world tropics (Cooper and Cooper 2004). There is at least a dozen species in Australia, with a couple of those found in rainforest. It is an unusual genus within the Lauraceae, for they do not grow as large trees like most species in the family, but as twiners. They are parasitic and usually lack any obvious leaves. On the stems develop tiny flowers and small round green to pale fruits.

Family Idiospermaceae/Calycanthaceae

The one species family Idiospermaceae is sometimes placed into the larger family Calycanthaceae. The species below was first discovered and named over 100 years ago. Then it was 'lost' and not rediscovered until 1971 when several cattle died from eating seeds on a property in Daintree. After rediscovery, it was recognized as being quite unusual, with primitive features, and thus re-classified into it's very own family, the Idiospermaceae. In more recent years, it has been the subject of research by Cairns JCU on aspects such as reproductive biology, and how the two isolated populations north and south of Cairns are evolving differently.

Idiot fruit or **Ribbonwood** ***Idiospermum australiense*** (fruit pictured below, right) Can be fairly big tree, trunk covered in many small bumps. Drops a large brown baseball-like fruit, frequently seen split into segments (seedling larger picture below, near Marrdja boardwalk).

Order Piperales

Family Aristolochiaceae

This family contains mainly climbers in the Australian tropical rainforest, often high in the canopy.

Dutchman's' Pipe
Aristolochia spp.

The species in this main genus are called 'Dutchman's Pipe' due to the shape of their rather grotesque looking, and apparently odorous, sock-like, deep cup flowers. The leaves of the plants here tend to be longish and heart-shaped with distinctive rounded lobed bases either side of the stem (pictured at right). The fruit are ribbed capsules (pictured right) that open to release small, angular, sometimes winged, seeds. The leaves of some species are food for the caterpillars of the Birdwing Butterflies.

Family Piperaceae

Peppers
Piper spp.

This is a huge group of thousands of species found throughout the tropics of the world. Most are vines, with leaves that smell peppery when crushed. They produce a collection of small, usually reddish fruits that are produced on the stem opposite the leaf. This genus of course provides many of the different peppers. There are several species in the Australian rainforests and most have vaguely heart-shaped leaves with distinctive venation that radiates from where the leave joins the stem.

Common Pepper *Piper caninum*. Most commonly seen species of native pepper. While most other native peppers have bright red fruit, this species is distinguished by it's rather duller, more orange coloured fruits (pictured left at Daintree Discovery Centre).

(Monocots)

Order Alismatales

Family Araceae

The species in the family Araceae are usually climbers or herb-like plants. They are distinguished by the flower that is made up of the unique spadix and spathe, a flat leaf-like structure. Many of the aroids are well-known climbers in the Australian tropical rainforest.

Elephant Ears
Alocasia spp.

The native 'Elephant Ears' (pictured right) and the naturalized Taro (pictured below) are similarly massive-leafed ground plants. The first are found along tracks in the region, while the latter tends to be found near water, such as along the Daintree River.

Epipremnum spp.

This genus contains about 15 species, found in tropical Asia, Australia and the south-west Pacific (Cooper and Cooper 2004). There are 2 species climbing tree trunks in the Australian rainforest. The better known is the large lobed leaf **Native Monstera *Epipremnum pinnatum*** (pictured right), which is widespread in rainforest (and gardens) along the east coast of Australia, and up into tropical Asia and various Pacific islands.

Rhaphidophora spp.

This genus contains about 70 species, found in tropical Asia, Australia and the south-west Pacific (Cooper and Cooper 2004). There are 3 species climbing tree trunks in the Australian rainforest.

Thick-leaved Rhaphidophora *Rhaphidophora hayi/pachyphylla*. Leaves are small and distinctly pressed flat against the trunk (pictured right). Leaves may begin size of thumbnails at base of trunk, but as grows higher, leaves can become as large as dinner plates. Common along the Daintree rainforest boardwalks.

Pothos spp.

This genus contains about 50 species, found in Madagascar, tropical Asia, Australia and the south-west Pacific (Cooper and Cooper 2004). There are 2 species in the Australian rainforest. They are both climbers that produce fleshy red fruit (upper picture left). ***Pothos longipes*** is a distinctive local species with 'candle and flame' shaped leaves (larger picture left).

Order Asparagales

There has been much shuffling of species and families within this order in the last few decades, and there still remain some classification work to be done. In this, families will be grouped together.

Families Asparagaceae/Agavaceae/Dracaenaceae/Laxmanniaceae/Xanthorrhoeaceae

Members of these families provide various uses for humans, including an alcohol and food from the rhizome of several species, but it is probably most functional for the fibres for clothes and rope that many different species can provide. The family is of also well known for its ornamentals such as Agave, Dracaena and Yucca, many of which can tolerate the 'lazy' gardener and still quite happily survive without watering, with some species also providing spectacular panicles of flowers. And lastly, since ancient Greek times, the young shoots of a particular cultivated plant have been eaten that we call Asparagus.

Palm Lilies, 'Ti'
Cordyline spp.

There are some 20 species of Cordylines found throughout the countries that border the tropical Pacific. They are usually small shrubs, but sometimes grow to several meters tall, with a spindly trunk. They has long leaves with a prominent mid-rib, and venation that is almost parallel but runs at a slight outwards angle. They produce small pink or white tubular flowers that often have yellow tipped stamens. The flowers and eventual fruits grow on a raceme at the end of a stalk erupting from the top of the plant. The stalk may droop or hang to the side of the plant. The seeds are enclosed in berries that are usually bright colours, such as yellows and reds. Cordylines are often used as ornamental plants as they can grow well in harsh conditions, with different species withstanding salty conditions, or a lack of light or water. Some species were also spread throughout the Pacific due to their starchy rhizome being used as a food source (Warren *et al* 1999). There are 8 species in Australia, with most of these being found as understorey plants in rainforests or wetter sclerophyll forests. ***Cordyline cannifolia*** (pictured right) is the common species of North Queensland tropical rainforest.

Families Alliaceae/Amaryllidaceae

This group contains a range of ornamental plants, such as the various daffodils, and the 'crinum lilies' (which are in a different order from the 'true lilies'). More importantly, it includes many species that provide important vegetables and flavours for kitchens around the world, including chives, leek and onion (Heywood *et al* 2007)

Mangrove Lily *Crinum pedunculatum* (pictured right at Dubuji). Short but large leaved plant with long leaves erupting from ground. Produces a stalk with white spider-like flowers, giving rise to other common name of 'spider lily'. Found in amongst mangroves along Marrdja boardwalk, along the Daintree River and in the rainforest behind the beach at Cape Tribulation.

Family Boryaceae

This family contains only two genera, both are which are only found in within Australia (Heywood *et al* 2007)

Resurrection plant *Borya septentrionalis*. Tiny erect herb that usually goes unnoticed until a dry spell, when tiny needle-like leaves go orange-red and appear to die. After rain, plant comes back to life as leaves green up. Small head of dense flowers are held erect from plant. Lives on high rocky slopes of mountains, and can be seen from some distance when they are dying and going red (pictured below on granite slopes of Mt. Hemmant).

Family Orchidaceae

This is one of the biggest (and some would suggest most highly evolved) plant families. Depending on the classification, there may be as many as 20,000 species in the family (Heywood *et al* 2007). Unlike the terrestrial orchids of temperate ecosystems, most tropical orchids, including those in the Australian tropical rainforest, are epiphytes, growing high in the canopy. The flowers are often elaborate and bizarre (higher picture below); they can be colourful, intricately patterned, and have an uneven petal arrangement ('zygomorphic'), with twisting and elongation of the sepals and petals. The following fruits are usually small, hard, dull capsules (lower picture below), that release lots of tiny seeds. The often unusual flowers are of course well sought after by human 'orchid hunters' and obsessively cultivated, and the original flavour vanilla comes from an orchid seed pod.

Dendrobium spp.

There are a few different species in this genus commonly known as 'pencil orchids', as they grow long cylindrical leaves (pictured right). Different species can be observed growing in the canopy in a range of habitats, including mangroves, lowland rainforest and upland rainforest.

Bottlebrush orchid *Dendrobium smillieae*. Large and spectacular orchid that develops 'bottlebrush' like head of many tiny little flowers that bunch at end of a long stalk. Individual flowers often have pink tubes, delicate whitish-green outer petals, and black or green central petals (pictured left, at Crossroads Cafe). Popular flower in cultivation.

Golden Orchid *Dendrobium discolor* (pictured below). May be epiphytic (growing on trees), more commonly seen growing as a lithophyte (growing on rocks). Spectacular orchid, with large sprays of colourful flowers growing on long woody stalks. Stems may stretch out for a metre, making it biggest orchid in Australia. Flowers vary in colour depending on variety, but usually a gold colour (thus the common name). It grows in beach vegetation, mangroves, open forest and rain forest in tropical eastern Australia and New Guinea, and seems to do well in coastal situations.

Order Pandanales

Family Pandanaceae

This family has about 3 genera with some 800 species (deWit 1967). Members of the family include trees, shrubs or climbers, and they have long, strap like leaves, usually with spined edges. Many species grow unusual roots, such as stilts at the bases of the trunk, or 'brace roots', off the trunk or limbs. They grow mainly in forests and along coasts in the Old World tropics.

Climbing Pandan
Freycinetia spp.

There are about 175 species in this genus (Cooper and Cooper 2004). They are climbers that usually have long thin, serrated leaves. The small flowers are actually inside large, often colourful bracts. The musty smelling flowers attract bats (deWit 1967). The genus is named after the French Commander Louis-Claude de Freycinet who visited Australia in 1817 (van Raders 199-). There are 5 species in Australia, and all of them are found in the north-eastern rainforests (Cooper and Cooper 2004).

Four of the five Australian climbing pandans have long leaves pictured right, Daintree Discovery Centre), look fairly similar and have overlapping distributions. Most are also found in New Guinea. ***Freycinetia excelsa*** is also found down to the subtropical rainforest in northern NSW. ***Freycinetia perocostata*** has the largest of the fruits.

Freycinetia scandens. Most distinctive of the Australian rainforest species, for leaves are small, short and rounded in shape (pictured below). Found in tropical rainforest and monsoon forest along northern half of Australia's east coast and into New Guinea (Nicholson and Nicholson 2000).

Pandanus spp.

This is the main genus in the family with about 500- 600 species, found mainly in the old world tropics (deWit 1967). There are about 20-40 species in Australia (Hyland *et al* 2000).

Pandanus gemmifer (below). Distinctive species has very long slender branches. Main leaves sprout in clumps from very end of branches; there are also smaller clumps of leaves just below main clumps. Bright red aggregate fruits hang down from main terminal clumps of leaves. Tends to grow sporadically along freshwater courses in tropical rainforest of north-eastern Australia.

Scrub Breadfruit *Pandanus monticola*. May grow as small tree but often just squat shrub. Collection of flowers look like corn in the cob (pictured below at left). Fruit a spiky ball that ripens from green to red or orange (pictured below at right). Found in understorey of tropical rainforest in north eastern Queensland, where it often forms dense stands in swampy coastal forest. Common along Marrdja and Dubuji boardwalks in Daintree. Provides the leafy food for 'Peppermint Stick Insect' and thus leaves commonly have holes in them.

Order Liliales

Families Smilacaceae and Rhipogonaceae

These two families are now considered separately, however, they are sometimes placed together, and their habit and leaves are very similar. There are at least 300 species in the first group, and only half a dozen in the second. The Smilacaceae are found throughout the tropics of the world, and in many temperate areas, while the Rhipogonaceae are found only in Australia and New Guinea. In Australia, the smilax vines are probably more obvious in the subtropical forests, but there are 6 species in the tropical rainforest (Cooper and Cooper 2004).

Both groups grow as vigorous climbers, with the wiry, curling tendrils growing in pairs from the petiole at the base of the leaves. Their leaves are easy to recognize, as they usually have 3 to 5 prominent veins, with the Smilax leaves growing alternatively on the stem, while the Ripogonum are mostly opposite (Cooper and Cooper 2004). The fruits grow in umbels (pictured at right, Cape Tribulation beach) in the Smilacaceae, and in a spike or a panicle in the Rhipogonaceae. They can be annoying plants for the main stems are covered in small prickles (pictured at left below). Several smilax vines provide the original Sarsaparilla (Heywood *et al* 2007).

Wombat Berry
Eustrephus spp.

The placement of where this plant belongs has been in dispute, with the possibility of inclusion in no less than five different families in recent years! (Philesiaceae, Smilacaceae, Luzuriagaceae, Geitonoplesiaceae, Laxmanniaceae). There is only 1 species in the genus; ***Eustrephus latifolius*** (pictured left, Josephine Falls). This thin vine has smooth pointed leaves with parallel venation. The small flowers have 'hairy' pink petals. The following bright orange fruits split in 3 to reveal segmented black seeds with a small amount of edible white flesh on the side (Low 1991). It is found in forests and woodlands along the entire east coast of Australia, and up into New Guinea and South-east Asia.

(Commelinds)

Order Commelinales

Family Commelinaceae

There are some 700 species in this worldwide family, including the purple flowered 'Wandering Jew', a common small weed on the edge of the rainforest.

Pollia spp.

There are over 25 plants in this genus and they are found throughout the tropics of the world (Cooper and Cooper 2004). They are usually short, scrambling plants with long leaves. They have the unusual habit of the main plant growing along the ground and sending up leafy stems regularly. The result is a mass of stems and leaves spread over the rainforest floor in clearing, all belonging to the one huge plant (pictured right, at Cathedral Fig tree).

Family Flagellariaceae

SuppleJack, Whip Vine ***Flagellaria indica***. Fast and vigorous climber with round, cane-like stem and long leaves. Long leaves are unusual as tips have been modified into tendrils to enable plant to curl its way through canopy. Found in mangroves and forests in eastern Australia and wetter tropical areas of Old world tropics. Common along all the Daintree boardwalks.

Order Arecales

Palms
Family Arecaceae

With at least 2,500 species, the palms are one of the biggest and most recognizable of plant groups. The tropical rainforests have 37 of Australia's 62 species (Cooper and Cooper 2004).

Archontophoenix spp.

This is a genus of half a dozen species found only in wetter habitats along the east coast of Australia.

Alexander Palm *Archontophoenix alexandrae* (pictured right at Lake Barrine). Tall, slim palm. Produces bunches of cream flowers. Followed by small bright red fruits that hang from below large green crownshaft. Commonly found planted in tropical gardens where known as 'King Palm'. Natural distribution only in swampy and water edge rainforests in eastern tropical Queensland.

Normanbya spp.

Black Palm *Normanbya normanbyi*. Tall slim trunk. Despite common name, more often whitish in colour when viewed from rainforest floor. Often penetrate canopy above as an emergent. Huge leaves are made up of alternative leaflets, paler underneath, appear to have 'chewed' ends. The fruits, which can often be found on the rainforest floor, are large, bright red with juicy and fibrous flesh, edible in small quantities. Only found in lowland tropical rainforests of Daintree, but within this tiny range is locally common. Easily seen around Daintree boardwalks. Resonant timber used by Aborigines to make musical clap sticks.

Cocos spp.

Coconut Palm *Cocos nucifera* (pictured right). Usually grows as a large tree, with slim trunk that often leans out towards or over sea. Top of tree is made up of large leaves that arch over their length. Produces the large familiar fruits which turn from green to brown as ripen. Fruit is made up of a smoother outer shell which contains another hard layer inside; this is the hairier husk that most people are familiar with. Natural distribution of coconut palm is issue of contention, for not only is it a naturally effective disperser but it has also been spread extensively by humans. Thus, for hundred of years has been common in many parts of tropical Indo-Pacific. Today, this 'exotic' plant is common along edge of coastal tropical rainforest.

Fan Palm
Licuala spp.

This genus has over a hundred species found mainly in the tropical forests of the Pacific Islands and Asia. There is 1 species endemic to Australia.

Fan Palm *Licuala ramsayi*. One of prettiest and striking of local palms. Endemic to lower swampier areas of north-eastern Australia. Has biggest leaf of any plant in the Australian rainforest (pictured below at Dubuji boardwalk). Produces small red fruits. Common along Cape Tribulation road and occurs in rustling stands along Marrdja and Dubuji boardwalks. There are also stands near Kuranda. Best place to see them is on the Licuala Fan Palm walk at Mission Beach.

Walking Stick Palms
Linospadix spp.

The genus of 'Walking Stick Palms' contains some 10 species. They are found in the tropical and subtropical rainforests of New Guinea and Australia, with 5 species in the latter, and of those, 4 occur in the tropical rainforest (Hyland *et al* 2002), and 1 in subtropical rainforest. They are fairly short, slender palms, usually with long narrow leaves with a fishtail-like terminal leaf. They produce small, colourful red or orange fruits that resemble tic-tacs along long spikes that erupt from the top of the plant. They get their common name of 'walking stick' as the southern species from the sub-tropical rainforests in northern NSW and southern Queensland (***Linospadix monostachya***) was used for walking sticks.

Walking Stick Palm *Linospadix microcarya.* The specific epithet refers to the small fruits; they are small, round dark orange and red fruits (pictured left, Josephine Falls). Grows in tropical rainforest from Innisfail to Mt. Lewis, from sea level to the mountains (Cooper and Cooper 2004).

Minor Walking Stick Palm *Linospadix minor.* Orange fruits (pictured left, Daintree) to red fruits (pictured above) fruits with large and long seeds relative to fruit; the little flesh they do have is edible. Grows in tropical rainforest from sea level to uplands in North Queensland and Cape York (Hyland *et al* 2002) also New Guinea (Jones 1995). Common in the lowlands, easy to see around the Daintree boardwalks

Linospadix palmeriana and Linospadix apetiolata both have red fruit and occur at higher altitudes (Jones 1995).

Calamus spp.

There are almost 400 species (Jones 1995) in this huge genus of palms. Some species are used for rattan furniture. There are half a dozen species found in the Australian rainforests where they are known locally as 'Wait-a-while'.

They are one of the most notorious of Australian rainforest plants, with nasty spines on the leaves, stalk and on the long, whip-like flagella, enabling the plant to climb into the competitive rainforest canopy. These spines often catch onto people travelling through the rainforest, and extraction can be slow and painful, so they have various derogatory nicknames, such as 'wait-a-while' and 'lawyer cane'.

The nasty tendrils are actually modified inflorescence stalks; that is, they are the stems for flowers. Every now and then the palm flowers and is then followed by clusters of small fruit. For all their nuisance to visitor of the rainforest, these palms were handy for the native peoples of Australia and tropical Asia. The fruits were eaten, the spines were used for catching fish, and the tough and flexible stalks used for climbing and building.

Common species around Australian tropical forests include; **Hairy Mary *Calamus australis***, and **Viscous Hairy Mary *Calamus radicalis***, both with fine spines, latter with many more leaflets. **Yellow Lawyer Cane/ Wait-a-while *Calamus motii*** (pictured right, Marrdja boardwalk) has much thicker yellow hooks (thus the common name) on midrib of leaves and sheath of plant; the latter are so large and stiff that when flicked they make 'plunk' sound. The species name *mottii* is a local aboriginal word.

Fishtail Wait-a-while *Calamus caryotoides* (pictured left, Atherton Tablelands). Wider leaves. Distinctive fish tail shape in terminal leaves. Seems to be more common on basalt soils.

Order Zingiberales

This order contains large-leafed and sometimes quite large herb-like plants. The most famous members are the bananas and the gingers.

Bananas
Family Musidae

Bananas
Musa spp.

There are about 35 true species of bananas and of course many different varieties. They are naturally found in the old world tropics, from Africa, through Asia, to Australia and some Pacific islands. In the wild, bananas are very much 'bat plants', pollinated by bats that take nectar from the huge hanging flowers. Bats also disperse the seeds when they eat the fruit.

There are two species of banana that are found naturally in Australia, and they are both found in the tropical rainforest of north-eastern Australia. ***Musa jackeyi*** has smaller orange-red fruits. ***Musa banksii*** looks fairly similar in general form to the cultivated banana, except that the fruits in the native species are held erect rather than hanging down like the cultivars. For more information, refer to the later section on agriculture.

Family Costaceae

This family contains over 120 species of 'gingers'. They are often placed with the 'true' gingers in the family Zingiberaceae to which they are closely related, however, there are differences. They tend to have shorter, wider leaves that are arranged spirally, join to the main stem with a closed and tubular sheath, and are not aromatic. The flowers in this family are produced as an inflorescence on a spike that looks like a cone, and this is usually located on the end of the pseudo-stem, or on a separate shorter stem from the ground; this is the attractive pineapple-like bract which makes species in this family such valued garden plants. The flowers usually have an enlarged lower 'lip', (in fact, these are modified stamens rather than true petals) often with a wavy margin. These plants are found mainly in the tropics of Africa and South America, but are also found in tropical Asia and Australia (Larsen *et al* 1999).

Costus spp.

There are 2 species in this genus found in Australia and they are both found in the tropical rainforests. One species is an exotic that has become naturalized, while the other is a native.

Spiral Ginger *Costus potierae*. Small shrub with leaves growing from spiralling stem. Produces white (with some yellow) flowers from bright red bract at end of the pseudo-stem. Found only around Cape York and close islands, and north-eastern Queensland in Australia, being quite common in Tully region. Within this distribution, is normally found in gaps and on edge of rainforest, mainly at sea level (Hyland *et al* 1999).

Backscratcher gingers
Tapeinchilos spp.

There are some 15 species of these striking gingers (Warren *et al* 1999). Their most distinctive feature are the 'backscratchers'; spectacular and bright red, plastic looking bracts that bear the smaller flowers and then later hold the seeds. They are found mainly in tropical Asia, with the following species also found in Australia.

Backscratcher Ginger *Tapeinochilos queenslandiae/anannassae*. Shrub with leaves held on slightly curling stems. Develops a large red plastic-like 'backscratcher' on top of straight tall stalk, from which yellow flowers grow. This striking bract also gives plant its other common names of 'Pineapple Ginger' and 'Torch Ginger'. Grows in tropical rainforests of north-east Queensland New Guinea (Nicholson and Nicholson 1995), and (depending on your classification) is also found in South-east Asia.

Family Zingiberaceae

This family contains the 'true' gingers. They have large leaves that are arranged in two rows along the stems. They often have aromatic rhizomes and leaves. Many species have showy flowers that have a fat 'lower lip' which looks rather like an orchid. Of course, this family includes the species from which the flavouring ginger is extracted.

Cardamon Gingers
Hornstedtia spp.

Native Cardamon or **Scott's Ginger *Hornstedtia scottiana***. Large ginger with huge leaves and swollen bases. Bright red bracts develop at ground level (pictured right, Mossman River), from which tiny pink and white flowers grow. Later, round red fruits develop inside these fading bracts, immersed within a slimy, snot-like substance. Ground birds such as cassowaries and brush turkeys eat the fruits (Nicholson and Nicholson 2000). Often called 'Native Cardamom' as seeds have similar flavour to related Asian plant. Seeds and pulp of this plant are edible, and leaves were used for wrapping food (Nicholson and Nicholson 2000). Usually found on edge and in gaps of tropical rainforest in northern Queensland, New Guinea and some other Pacific islands (Nicholson and Nicholson 2000).

(New Guinea) Gingers
Psychanthus spp.

This genus of over 20 species is found mainly in New Guinea and surrounding islands, with 1 species found in Australia.

Psychanthus racemiger. Medium sized ginger that develops elongated orange fruits from leaf stalks (pictured, at Daintree Hideaway).

Alpinia spp.

These gingers are found mainly in tropical Asia and the Pacific. There are 4 species native to Australia. They often have pretty flowers with pouting lower 'lips' growing from the end of the main pseudo-stem between the leaves.

Snowflake Ginger, Pleated Ginger *Alpinia arctiflora* (pictured right, Possum Valley). Attractive ginger with wavy margins to leaves. Small hairs on underside of leaf produce soft, velvety feel (making it a great outdoor toilet paper). Other common name of 'snowflake ginger' alludes to pretty white flowers produced at end of main pseudo-stems; have yellow centre and are reported by Nicholson and Nicholson (1995) to be biggest flower of any Australian ginger. Followed by an unattractive collection of fruits that at first green, then turn into what looks like small brownish-gray bananas. The many small seeds and their flaky white surroundings can be nibbled, but generally not considered edible. Found in gaps on edge of rainforest in North Queensland, seems more common at mid and higher altitudes.

Narrow leaved Ginger, Stalkless Ginger *Alpinia modesta*. Small ginger of about metre in height. As common names suggest, long leaves are narrow and have no stalk, attaching directly to main (pseudo-) stem. Reddish tinge often present on edge of leaves or along stem. Flowers quite pretty, with puffy white lips, often with tinge of pink or red. Fruits are small blue berries. Only found in rainforests of North Queensland, but widespread within that range.

Blue Ginger *Alpinia caeulea*. Medium sized ginger that may develop purplish tinge to underside of leaves (Cooper and Cooper 1994). May have slight lemon scent when crushed (Roberts *et al* 1995). Produces white flowers which are followed by (as common name suggests) small round blue fruits. Once crisp flesh is taken off, the flaky white flesh can be eaten and seeds sucked, or even chewed, for full strong flavour. Fruits were popular with rainforest aborigines, and they were tracked by their trail of spat out seeds (Low 1991). Aborigines also used leaves for wrapping food during cooking and ate tips of rhizomes that were ginger in taste (Roberts *et al* 1995, Low 1995).

(Eudicots)

Order Proteales

Family Proteaceae

The family Proteaceae contains some 1600 species in about 77 genera (Cooper and Cooper 2004). Species can be found in the tropical Americas, Asia, Africa and Australia. However, the family is best known for the large and colourful flowers of Banksias and Grevilleas from the temperate woodlands of south-western and south-eastern Australia, and Proteas from the heathlands of South Africa, and it is these areas where there is the highest diversity.

Their distribution reaffirms the long ago Gondwanan connections between southern Africa, South America and Australia/New Guinea. Interestingly, the genus *Gevuina* has only 3 species, one each in South America, New Guinea and the Australian tropical rainforest. About two thirds of the species of the family are endemic to Australia. And while most species, especially the better known ones, are known for their adaptations to arid and fire effected habitats, many of the ancestral species are found in the tropical rainforests.

With their spectacular flowers, many of the Proteaceae family are used as ornamentals, including rainforest species such as the Firewheel Tree and various Silky Oaks. One member of the family has also been cultivated by the U.S.A., although it was originally from the rainforest of eastern Australia; the Macadamia nut.

Life cycle - flowers, fruits, seeds, leaves and trees

The famous, showy flowers of most species in this family are usually collections of lots of tiny flowers (an inflorescence) along a spike. Many of these large collections of flowers produce dripping amounts of nectar. This entices many animals, and thus the plants are pollinated by a range of animals including insects, honeyeaters, lorikeets and mammals. In many species the petals/sepals are connected into a hunched tube, the stamens are short and closely attached to the sepals, and the most obvious and colourful part of the flower is the long, protruding style.

Thus, in many groups, including *Grevillea* and *Darlingia*, it is the female part of the flower that provides the colour rather than the petals (as in most flowers), or the stamens (as in the superficially similar bottlebrushes and paperbarks of the family Myrtaceae). Before the flower has opened, the pollen has been transferred to the end of the long style to what is referred to as the "pollen presenter". After this pollen has been dispersed, the stigma then becomes receptive to receive pollen from other flowers (Clarke and Lee 1997).

The following fruits can vary from small fleshy fruits to large, hard and woody follicles. The fleshy fruits may have a persistent style still attached. The woody fruits open up to let out their light, wind dispersed seeds (pictured right).

Many species have seedlings that start out with a distinctive butterfly shape (pictured left). From this humble beginning they grow into some of the biggest trees of the Australian tropical rainforest.

The origin of the name of the family Proteaceae (protean) is an allusion to the ability to change shape, and indeed many rainforest species change the shape of their leaves as they grow; for example, they may begin as shrubs with lobed leaves and then grow into large trees with entire, simple leaves. As mature trees, many of the giant Australian rainforest Proteaceae were well sought after timbers. Today, many of those plants are still known by common names given to them by timber-getters; thus there are many different types of 'silky oaks'.

Proteaceae in the Australian tropical rainforests

The Proteaceae are a very important component in the Australian tropical rainforests of Australia. Of the 1,100 species of Proteaceae found in Australia, more than 50 species can be found in the tropical rainforest. The remarkable thing about this is that there are 26 genera; that is the tropical rainforests have a high diversity of different *types* of Proteaceae. Many of the genera have a only a few species, or a monotypic (just one species). As with many other species in the rainforest, they are endemic to the region and thus rare plants. The tropical rainforest of North-east Queensland is also home to what are presumed to be many of the most primitive members of the family. From these species they are thought to have then evolved into the Proteaceae that we are more familiar with in the woodlands, such as the Grevillea and the Banksia.

Athertonia spp.

There is only 1 species in this genus, and it is endemic to the tropical rainforests of North Queensland.

Atherton Oak *Athertonia diversifolia*. Tree to 30 metres, may develop buttress roots. Leaves are variable (as suggested by specific epithet *diversifolia*), being large and lobed when young, develop into simple leaves as tree grows; often rusty coloured underneath, especially on central veins and new growing leaves. Long racemes of flowers are cream or brown in colour. Fruit is slim, oval-shaped and ripens from green to bright blue; woody covering underneath the flesh and around seed has very distinctive pits (pictured). The single seed inside is apparently delicious (Nicholson and Nicholson 1988), eaten by various ground mammals; there is often chewed hole (pictured) from White-tailed Rat. Found in tropical rainforest of north-eastern Queensland at a range of altitudes.

Macadamia spp.

There are 8 species in this genus. All but 1 species is endemic to the rainforests of the east coast of Australia; the other species is found in the islands of New Caledonia (Cooper and Cooper 2004). The name *Macadamia* is named after Victorian chemist John Macadam. The genus is notable for providing the only example (so far) of a native Australian plant that has produced an international agricultural product. Of course, the Aboriginal people of the rainforest have long consumed the fruits of the Macadamia trees. The species that provide the edible nuts are ***Macadamia integrifolia*** and ***Macadamia tetraphylla*** both from the subtropical forests of southern Queensland and northern NSW (Jones 1986). There are 3 species in the tropical rainforest of North Queensland.

Grevillea spp.

There are at least 360 species of Grevillea, and they are found throughout Melanesia and Australia. However, most of the worlds species are endemic to Australia, where they are mainly found in the woodlands and heathlands of the south-east and south-west of the continent. There are only 2 species commonly found in the tropical rainforest.

Grevillea baileyana and ***Grevillea hilliana.*** Both begin as juveniles with lobed leaves, develop simple leaves as mature. Leaves in both distinctively coloured underneath; in former species are almost golden (pictured right), and in latter, coloured silver. Both develop small round brown follicle fruits that open to reveal coin-shaped seeds.

Rose Silky Oaks, Brown Silky Oaks *Darlingia* spp.

There are only 2 species in this genus, and they are only found in the rainforests of north-east Queensland. Both ***Darlingia darlingiana*** and ***Darlingia ferruginea*** can grow to huge trees and may develop buttress roots. Both species commonly have lobed as well as entire leaves, but the latter species has rusty coloured hairs underneath the leaves, thus the scientific name. Both species produce spectacular displays of cream coloured flowers (pictured left). This is followed by light brown, finger length oblong woody fruits that open (pictured below, Cairns Lakes) to release winged seeds.

Stenocarpus spp.

There are some 25 species in this genus and they are found from New Guinea, through Melanesia, to Australia. Many species are known as 'wheel of fire' due to the circular arrangement and bright red colour of the flowers (right). Some of the flowers in the genus ***Alloxylon*** are similar, but tend to be more erect. There are about half a dozen species found in the Australian tropical rainforest, and they seem more common at higher altitudes, such as the best known species ***Stenocarpus sinuatus***, which is planted extensively in gardens.

Austromuellera spp.

There are only 2 species in this genus, and they are both endemic to the tropical rainforests of north-eastern Queensland. Both **Mueller's Silky Oak *Austromuellera trinerva*** and ***Austromuellera valida*** grow as trees. They produce long hanging racemes of flowers from the branches, followed by large woody fruits that open to release winged seeds (pictured below, fruit of ***Austromuellera trinerva***, Mardja boardwalk).

Musgravea spp.

This genus only contains 2 species, and they are both endemic to the rainforest of North-eastern Queensland (Cooper and Cooper 2004). Both **Briar Silky Oak *Musgravea heterophylla*** and **Crater Silky Oak *Musgravea stenostachya*** grow to large trees, and may develop buttress roots. They develop finger length, oblong, pointed end woody follicles, covered in rusty or grey hairs. While still sitting erect from the branches they split open to release attractively brown, flat seeds (*Musgravea heterophylla* seeds pictured left, Mossman Gorge).

Cardwellia spp.

There is only 1 species in this genus. The genus was named after the secretary for the colonies, Edward Cardwell (Cooper and Cooper 2004). Although the trunk is not always straight, this species has been one of most important trees of the Queensland timber industry.

Northern Silky Oak *Cardwellia sublimis*. Can be one of the dominant trees in Australian tropical rainforest, often standing well above canopy. From this height the large, erect, wooden hoof-shaped fruits release their winged, wind-borne seeds. Follicle is quite hard, and after it has fallen onto forest floor, may remain for some time; this is where most people first notice this species. Restricted to tropical rainforest of north east Queensland.

Order Ranunculales

Family Menispermaceae

There are over 500 species in this family found in the tropics throughout the world. In the Australian rainforest, the family consists mainly of climbers. The leaves are simple and alternate and are often heart shaped. The flowers are small. The following fruits are often colourful, and because the style/stigma is slightly off centre, they have a lopsided look to them. The small seeds inside are odd looking, often being kidney bean or horse-shoe shaped, and sometimes ribbed.

Snake Vine or Tape Vine *Stephania japonica*. Thin vine with peltate leaves. Usually only noticed when female plant produces its bunch of dullish red fruits on main stalk, often quite low down in the canopy. Widespread distribution from tropical Australia, through tropical Asia to Japan.

Pycnarrhena spp.

There are 9 species in this genus and they are found from Australia, through tropical Australia to China (Cooper and Cooper 2004). Two of these are found in the tropical rainforests of north-eastern Queensland and Cape York. They are both vines with simple, long leaves. They develop 'lopsided' orange to red fruit with a single kidney shaped seed (pictured below is ***Pycnarrhena novoguineensis***, at Cape Tribulation).

(Core Eudicots)

Family Dilleniaceae

The family Dilleniaceae is made up of some 400 species of vines and trees; in the Australian tropical rainforest the species in this family tend to have the heavily veined 'crinkle cut chip' look to the leaves, and many produce large yellow flowers.

Dillenia spp.

There are some 60 species of *Dillenia* found in tropical Asia and the Pacific, with only 1 species native to Australia.

Red Beech *Dillenia alata*. Distinctive flaky red bark (larger picture at right), large yellow flowers (lower picture right), and red fruits that open up like a flower to reveal the seeds (pictured below, at Cape Tribulation). Common in coastal forest in Queensland and NT, and up into New Guinea (Cooper and Cooper 2004). Obvious at Cape Tribulation picnic area.

Fire vine
Tetrecera spp.

The 40 species in this genus are found all around the tropics of the world, with 2 species found in the Australian tropical rainforests. These vines often have the 'crinkle cut' look to the leaves like many other species in this family. They are referred to as 'fire vines' because of the sand-papery feel to the leaves and branches, and which certain people may be more sensitive to. They have very short-lived white fragrant flowers. The following fruits are brown or green capsules with distinctive points to them. They open up like flowers to reveal strange frilled red arils with small black seeds inside.

Order Santalales

Mistletoes

Families Loranthaceae, Santalaceae and Viscaceae

The common name 'mistletoe' refers to many species in several different, but closely related, families. The different groups dominate in different regions; the Viscaceae are more of a northern hemisphere family, centred in Asia and notable in European and American Christmas culture (Heywood *et al* 2007), while the Santalaceae are widespread in more tropical and drier regions around the world, and the Loranthaceae are more prominent in the tropics and the southern hemisphere, including Australia; these are the main mistletoes seen in the rainforest.

Most mistletoes are plants that live in and around the branches of trees and shrubs. They appear as climbers or even thick aerial bushes within the canopy. They are parasites, clinging to their host plants with suckers running around and sometimes into the bark. Mistletoe berries have evolved in an interesting way; their seeds are surrounded by a very sticky substance that forces the birds eating the fruit to wipe their beaks or bottoms onto branches to get rid of it, and thus potentially transplanting the seed to the canopy of a new host. In Australia, there is even a bird, the aptly named Mistletoebird (actually a modified flowerpecker), that specializes in eating the fruits, and subsequently dispersing the seeds. There are interesting variations on the parasitic mistletoe lifestyle; there are some species that parasitise the roots of their host plant, there are tree-sized species, and within the family Viscaceae there are even mistletoes that parasitise other mistletoes! Australia has a high diversity of mistletoe plants, and in the rainforest there are representatives from all three families.

Mistletoe species vary much in detail. Those in the Viscaceae are usually runners or shrubs, often with small or no leaves, while species in Loranthaceae tend to be shrubbier in appearance with more obvious leaves. In Australia, many of the mistletoes of the Loranthacae grow very similar foliage to their hosts. Species in the Santalaceae range from climbers to shrub and tree size parasites, such as the Sandalwoods. Species in the Viscaceae often only produce tiny flowers, and small fruits, while those in Loranthaceae tend to have brightly colourful, showy flowers (pictured below, Cape Tribulation) and colourful berries, and it is not uncommon to see a huge patch of red or orange flowers in the canopy due to mistletoe 'infection'.

(Rosids)

Order Myrtales

Family Myrtaceae

The family Myrtaceae is arguably the most important family on the Australian continent. And while they are not as obviously dominant in the tropical rainforests as they are in the drier woodlands, the group includes many different species in this habitat.

The most useful diagnostic feature of this family are the leaves and their obvious oil glands, which appear as light dots when the leaf is held up to bright light. The flowers in the Myrtaceae family are usually quite showy, but not in the usual way; instead of colourful petals, there is instead a seemingly unlimited number of stamens, giving the flower the appearance of a brush (pictured above).

Botanists often divide the family into two different sub-families. One group are the woody fruited Myrtles. They are dominant in drier parts of our continent and include the classic Australian genera such as *Eucalyptus*, *Melaleuca* and *Callistemon*. These groups are only represented by a few species in the rainforest. The other group are the soft-fruited myrtles. They are usually associated with wetter environments and found mainly in the tropical rainforests of the Americas, Australia and parts of the Pacific. It includes such genera as *Syzygium, Acmena* and *Eugenia*, and are commonly known as the 'Apples', Satin-ashes and Lilly-pillies. In Australia they were given their confusing and misleading common names by the early European timber-getters, as they reminded them of trees or timber properties from back home. For example, the name 'satin-ash' was given to some trees as the timber was midway between the properties of the satin-woods and the ashes (Boland *et al* 1999). They are often huge trees, and are found mainly in wetter habitats, such as rainforest. They often have showy flowers and produce fleshy fruits. This fruit is what gives them the common name of 'apple', but they are not related to the regular supermarket apple.

Apples, Satin-ashes and Lily-pillies
Syzygium spp. (and former *Acmena* spp.)

There are about 500 species in this genus, and they are found throughout the old world tropics of the world, usually in wetter habitats. Of the 60 or species in Australia, about 50 occur in the tropical rainforests of the east coast (Cooper and Cooper 2004). *Syzygium* are mainly trees, ranging in size from stunted bushes, to towering rainforest giants. They usually have simple, opposite leaves that are often glossy and rounded.

The name derives from *syzygos*, which refers to the flower segments being fused together to form a cap when a bud, and are shed as the flowers open (Cooper and Cooper 1994). The flowers are often showy, and dominated by many colourful shaving brush like stamens. They develop fleshy fruits that are amongst the most colourful fruits of the Australian rainforest. These colourful fruits provide food for many animals, and many species are edible. A tropical Asian species provides clove.

Water Gum, Grey Satin-ash *Syzygium gustavioides*. Can be large tree. Very distinctive dark; brown trunk, often develop 'flying' buttress roots (pictured right, Marrdja boardwalk), sometimes large enough to walk through. Distinctive fruit; large and dirty green, thin flesh, persistent calyx ring that looks like crater ring, varies in size. Tree found in better developed rainforest and also swampier areas, thus the name 'water gum' (not to be confused with the riverside Water Gum), in north-eastern Queensland only.

Paperbark Satin-ash ***Syzygium papyraceum***. One of few distinctive trunks in rainforest, where most trees look the same. Seems to usually grow as fairly slim tree, with very orange to red flaky bark, which is even brighter in wet conditions (pictured right at Possum Valley). Common name and specific epithet refers to this papery bark. Flowers fragrant and coloured white to purple. Elongate rounded fruits distinctively purple, size of large marbles. Found at higher altitudes in North-eastern Queensland (Cooper and Cooper 2004).

Kuranda Satin-ash ***Syzygium kuranda***. Can be large tree, may develop buttress roots. Produces dull coloured fruit, an orange/pinkish brown (pictured left), with very thin calyx lobes. Found in rainforest at most altitudes throughout North-east Queensland.

Cassowary Satinash, Red Watergum ***Acmena graveolens*****.** Can be large tree with elaborate buttress roots. Distinctive, unusual, large pink large. Produces a large pink fruit, with distinct ring around flatter end, skin is hard and edible flesh is thin (pictured below, Cape Tribulation Beach House). Found in lower altitude rainforests in north-east Queensland

White Apple *Syzygium forte*. Large tree, especially in very wet situations; gullies in well developed rainforests. May develop buttressing, bark is flaky. Leaves are opposite, large, glossy, smooth and rounded; begin bronze colour, grow to light green, become darker green, remaining paler underneath. Flowers are impressive large sprays, densely packed cream coloured stamens, usually grow erect from end of branches. Small to large round very white fruits (pictured left, at Mossman Gorge), calyx ring with tiny lobes, sometimes a short string; can fruit prolifically, littering ground underneath. Fruits edible, tasting crisp. Found in coastal Queensland, N.T, New Guinea, in various types of rainforests, manly in coastal areas, as fairly salt and wind resistant (Jones 1986).

Bumpy Satinashes

These two noticeable trees are restricted to the tropical rainforests of north-east Australia. They are noticeable because they are such great examples of 'cauliflory', with the showy, stamen rich flowers, and then the large, round, whitish fruits, erupting from the trunk. Even when not flowering or fruiting, the tree's bumps are distinctive in an ecosystem where all the tree trunks look very similar. Like many other Syzygiums, the fruits are rather strange, but not offensive, to taste.

Bumpy Satin-ash *Syzygium cormiflorum*. Fruits of this species are large, round and white, often with a pinkish tinge in contrast to species below, and have a very puckering calyx. This well known species is widespread throughout the wet tropics of North Queensland.

Daintree Satin-ash *Syzygium monospermum*. Fruits are round, large, very white, slightly glossy, slightly ribbed, brown bruising, and less puckered than above species (pictured below). The tree appears to have a special relationship with black ants *Iridomyrmex gilberti* that live inside the hollow bumps from where flowers and fruit sprout (Nicholson and Nicholson 2000). Restricted mainly to lowland tropical rainforest of North Queensland between Julatten and Cape Tribulation. Examples can be seen along Marrdja boardwalk (pictured right).

Archirhodomyrtus spp.

There are 10 species in this genus, and they are found in New Guinea, New Caledonia and Australia, where there is just 1 species; ***Archirhodomyrtus beckleri*** (pictured right, Alexandra Range lookout) a shrub found from NSW to northern Queensland.

Paperbarks
Melaleuca spp.

The Paperbarks are one of the most distinctive Australian plants, usually well known both to residents and many visitors alike. Melaleucas are often mistaken at first glance for Eucalyptus trees. They are in a separate genus, but in the same family, and many of the species share the familiar pendulous branches and drooping leaves. Other common family features are the aromatic, oil gland filled leaves and the flowers dominated by many colourful stamens (lower picture right). And as is obvious from the main common name, many species have peeling, papery bark (upper picture right). They are common in the woodlands and wetlands that are often closely adjacent or even merging with the tropical rainforest.

Pendas
Xanthostemon spp.

Among the few hard fruited genera in the Myrtaceae that are found mainly in the tropical rainforest (and some of the adjacent woodlands) is *Xanthostemon*. This name means "yellow flowers" and refers to the most obvious feature in this genus; the showy flowers are dominated by lots of yellow stamens. There are some 45 species in the tropical South Pacific, including a dozen species in Australia, with most found in the tropical rainforest (Cooper and Cooper 2004).

Pendas can be seen along some of the freshwater creeks in North Queensland, such as the Mossman River, Noah Creek and Bloomfield River and when they erupt with their spectacular yellow flowers it suddenly becomes obvious how locally common they really are. They have also been planted extensively along the roadsides around Cairns.

Family Melastomaceae

The family includes the genus *Tibouchina*, which provides the popular garden Lasiandras. There are about 5000 species in this neotropical centred family (Cooper and Cooper 2004). While most species are found in the rainforests of of South and Central America, the family is also distributed throughout the tropics of the world, including Australia. Locally, the Melastomids are usually shrubs, but they can also be climbers. The leaves are distinctive, with 3 to 5 strong central parallel veins. They often have bright and showy purple flowers. Melastomids often produce small vaguely raspberry-like fruits with edible flesh that can be squeezed out. There are several genera of melastomids in the Australian tropical rainforest, including native and introduced species. They are common on the edge of the rainforest, and they are frequently planted in native garden and re-vegetation areas (such as near Jindalba and Dubuji boardwalks).

Dissotis spp.

This is a genus of 100 species found throughout much of the tropics of the world, including 3 species in Australia. However, only 1 species is found in the Ausrtalian tropical rainforest, and it is not native. ***Dissotis rotundifolia*** is also commonly known as '**Spanish Shawl'** or '**Pink Lady**' (Cooper and Cooper 2004). It is a small creeper along the ground on the edge of the forest. It has the 3-veined leaves and pink/purple flowers (pictured right, Daintree Hideaway) typical of the family. It is originally from Africa, but is now found in the wild in Cape York, down to the Daintree River.

Melastoma spp.

This is a genus of over 20 species that are centred in India, through Asia, to New Guinea, Australia, and Pacific Islands (Cooper and Cooper 2004). The name *Melastoma* means 'dark mouth', and the plants are commonly called 'blue tongue', because eating the fruit stains the mouth.

Melastoma affine/malabathricum. Shrub with 5 prominent veined leaves and pink/purple flowers (pictured below, left). Produces small round fruit, of which inner purple flesh (pictured below, right) is edible and tasty. Found in a variety of habitats, including woodland and rainforest, from India, through Asia, New Guinea, Australia and Pacific islands (Cooper and Cooper 2004).

Medinella spp.

This is a genus of some 150 species that occur throughout the old world tropics, including just 1 species in Australia (Cooper and Cooper 2004).

Medinilla balls-headleyi. Grows as a climber. Produces shocking pink gum nut like fruits. The unfortunate species name is in fact after a person. Occurs in rainforest at lower altitudes in north-east Queensland.

Tristemma spp.

This is a genus of 14 species that occur naturally in tropical Africa and Madagascar (Cooper and Cooper 2004). One species has become naturalized in tropical Australia.

Tristemma mauritianum. Distinctive 3-5-veined leaves of the family, often has red stems. During the warmer parts of year, develops small light pink flowers. Followed by pinkish-red fruits, which often appear on plant at same time as flower; nice tasting flesh can be squeezed out and eaten. Originally from tropical Africa and Madagascar and was first collected in Mauritius, thus the scientific name (Cooper and Cooper 2004). Now found on edge of the tropical rainforest in North Queensland.

Family Combretaceae

There are over 500 species in 20 genera in this family (Hyland *et al* 2002). They are found throughout the tropics of the world, rarely in cooler areas, and are particularly well represented in tropical Africa. The family includes some edible species, such as the kernel of *Terminalia catappa*, ('Indian Almond'), as well as some ornamentals, such as the Rangoon Creeper *Quisqualis indica*. The fruits of one Australian outback species has been found to have 50 times more Vitamin C than oranges (Rohwer 2002). There are over 30 species found in Australia, with 20 of those found in the tropical rainforests (Cooper and Cooper 2004).

Terminalia spp.

This is a genus of about 200 species found throughout the tropics of the world, with some two dozen that occur naturally in tropical Australia (Hyland *et al* 2002). As suggested by the common name, the leaves in many species tend to be bunched at the terminal end of the branches. The leaves are mostly large, rounded and bigger towards the end. The flowers are usually small, without petals, cream in colour, and arranged along long spikes. The fruits are usually oval in shape, and are often hard'ish with a thin leathery flesh. They are usually coloured with variations of red, purple or blue and have hard to get at kernels inside. Several species are notable in coastal habitats, and there are also many species found in the tropical woodlands of northern Australia. Of a total of 24 species found in Australia, a dozen are found in the tropical rainforest (Cooper and Cooper 2004). The species below are **Beach Almond *Terminalia catappa*** (below, lower picture) and **Brown Damson *Terminalia arenicola*** (below, upper picture).

(Eurosids)
Order Celastrales
Family Celastraceae

This family has over 50 genera and over 1000 species (Cooper and Cooper 2004). They are found worldwide, but there are many more species in the tropics and subtropics. It seems to be a common feature of many of the plants found in the Australian tropics to have an orange colour underneath the first top layer of bark. The flowers are usually small and pale white or green, with 3-5 sepals and petals. The flowers usually have a distinctive fleshy or puffy looking disc in the middle. The stamens are often squat and stick out of the disc or are on the edge of the disc, and the style is usually short, with the ovary often partially buried in the disc. The following fruits vary in colour and form. The family provides a few ornamentals as well as *khat* tea which is drunk in the Middle-East (deWit 1965).

Salacia spp.

There are over 150 species in this genus, with 3 species found in the tropical rainforest of North Queensland, including the **Lolly Berry *Salacia chinensis***. This vine is named for the red, round fruits that look like lollies, and the sweet and tasty, sticky edible flesh. They seem to be most obvious on the edge of rainforest and monsoon forests, especially along waterways, from tropical North Queensland, through the Pacific islands, across south-east Asia and to India (Cooper and Cooper 2004).

Ivorywood
Siphonodon spp.

There are 7 species in this genus and they are found in tropical Asia, Australia and the Pacific. There are 2 species that can be found in the tropical rainforest of North Queensland, with the species below being the most common in the wetter habitats.

Siphonodon membranaceus. Can grow to be large tree. Orange layer underneath first layer of bark. Flowers are small and pale. Fruits look rather like small oranges when first seen on floor of rainforest (pictured right at Mossman Gorge). Easily crushed fruits have very distinctive flesh, which is sand-like in texture, but edible (although texture is bit off-putting). Many small angular seeds inside the flesh. Endemic to rainforests of North Queensland, where occur from Tully to Cooktown at most altitudes (Cooper and Cooper 2004); seem to most obvious in 'good quality' rainforest.

Order Cucurbitales

Family Cucurbitaceae

This family includes mainly climbing plants. They are found throughout the tropics and subtropics of the world. This family is important to humans, with many edible species, such as cucumbers, squashes, melons, gourds and pumpkins.

Striped Cucumber *Diplocyclos palmatus*. Thin-stemmed twiner. Fruits are very distinctive miniature watermelon-like fruits that start green with white blotched stripes, then ripen to bright red orange. Many small whitish seeds; are toxic. Very widespread, found in wetter forests and woodlands throughout tropical Australia, Asia, Pacific and Africa.

Gourds
Trichosanthes spp.

These plants grow as vines with palm-shaped leaves. They usually produce large orange to reddish fruits with many seeds inside. They are found mainly in tropical Asia, Australia and the Pacific. There are half a dozen species found in the rainforests of tropical Australia. Pictured is one of the larger fruited species, ***Trichosanthes pentaphylla***.

Order Fabales

The 'legumes' are a huge group of plants that are found all over the world in almost every habitat. They are sometimes placed in the one massive family, and at other times in separate families. Species in this group often have showy flowers and develop hard seed pods. The group includes many ornamentals (from golden wattles to golden shower trees), as well as species that provide us with food products, from soy bean to bean sprouts.

Family Caesalpiniaceae/ Subfamily Caesalpinioideae

This large family of legumes has over 2000 species in 150 genera (Hyland *et al* 2002). They are found all over the tropics of the world, with many extending to the subtropics and a few in temperate areas. The plants themselves can grow in a variety of forms, from vines to trees. The leaves are compound; that is, they are usually divided into small round leaflets, being pinnate or bipinnate. The flowers are irregular and can be quite elaborate, usually with well developed lateral or side petals, and thus appearing almost orchid-like. Being a legume, the resulting fruits are predominantly bean-like pods with seeds encased inside. Many of the plants in this family are used as ornamentals; for example, the 'Poinciana' *Delonix regia* and many others are common street tree in tropical urban areas such as Cairns.

Caesalpinia spp.

This is a group of some 100 species found throughout the tropics of the world. A dozen species extend into Australia, with about 8 of those found in tropical rainforest. The genus is named after an Italian botanist of the 16th century (Cooper and Cooper 2004).

The species found in the Australian tropical rainforest are all climbers. They have small sharp spines on the stems for climbing, and some also have spines on the seed pods. They have various local names reflecting their spiny nature; Prickly Poinciana, Prickle Vine, and Mother-in-Law vine. The leaves are compound with small leaflets, giving a misleading soft feathery appearance, concealing their somewhat prickly nature (pictured left, middle). They produce round flat pods that open to reveal the round or angular flattened seeds inside. The most widespread species is **Bonduc, Nicker Nute *Caesalpinia bonduc***, which is found on seashore vegetation around the tropics of world, including Australian lowland rainforest. A higher altitude species of the Atherton Tablelands is ***Caesalpinia robusta*** which develops a corky stem (larger picture at left, Red Cedar), a pod with sharp spines (lower picture, left) and purple, angular seeds.

Family Mimosaceae / Subfamily Mimosoideae

This group contains about 3000 species. Because of the genus Acacia, they are one of the most dominant families in the tropical woodlands of Africa, the Americas and Australia. Additionally, they also are an important group of plants in the tropical rainforest.

Shy weed *Mimosa pudica*. Small prostrate scrambling plant. Has the classic features of the mimosa, such as the fine feathery leaves and purple flower balls. Stems also often have spines. Most distinctively, leaves fold up when touched, thus giving plants it's common names of 'touch-me-not', 'shy weed, 'sensitive weed', etc. (The left photograph shows the weed before being touched, while the right photograph shows the same leaves after they have been touched and have closed a few seconds later). Is indeed a weed, originally from tropical Americas, but now found throughout tropics of world. Can be very common along roadsides in coastal lowlands of North Queensland.

Rain Tree *Samanea saman*. Large tree with spreading crown, resulting in a silhouette rather like an Acacia on African plain. Pink powder puff-like flowers. Called 'rain tree' for several reasons; the compound leaflets open and close slightly depending on humidity, and thus potential rain, also; there is a cicada that lives in the crown, and if present in large numbers the secretion of the insect as if feeds on the tree contributes continuous fine 'rain' underneath the crown (Rohwer 2002). Originally from South America, has been planted widely in tropical Asia and India, as well as north-east Queensland. Commonly planted by the Chinese in early settler days to protect crops, and they remain on the edge of the rainforest in many places, such as along the Daintree River (pictured below). Best known and most spectacular specimens are those either side of the highway exiting the north side of Mossman.

Wattles
Acacia spp.

The are more species of Acacia than any other genus of plants in Australia, with somewhere between 700 to 900 species found on the continent, of a total of 1,000 to 1,500 species worldwide. The plants themselves are mainly small bushes or large trees. They have either attractive bipinnate leaves or leaf-like "phyllodes", which are actually modified stalks that look and also perform the photosynthetic role of leaves. These phyllodes are usually long and hang vertically, and do not have true venation, but have parallel lines running along their length. The flowers of Acacias are distinctive; usually yellow and fluffy, either in cylindrical spikes or many single heads. The following fruits are hard pods, in a huge variety of interesting shapes and sizes. The seeds are usually attached to the inside of the seed pod by a strip of flesh called a 'funicle', which is sometimes colourful. While Acacias are more obviously dominate in the drier environments of Australia, there are several species that occur in the rainforest, especially in re-growth forest and on the edge; the places where people tend to be, and thus they are commonly encountered.

Acacia mangium. This species can grow to a large tree in the forest, and develop buttress roots. Pods are collection of small tightly twisting curls. Black seeds are initially attached with yellow to orange funicles, but after these drop, the 'springs' may appear on the ground for some time afterwards (pictured right). Occurs on edge of various habitats, including woodlands, mangroves, and rainforest, as well as inside well developed rainforest, especially re-growth. Also found up into tropical south-east Asia (Hyland *et al* 2002).

Black Wattles, **Salwoods**, **Earpod Wattles**, ***Acacia aulacocarpa***, ***Acacia auriculiformis*** and ***Acacia celsa***. Series of closely related and similar Acacias. Known as 'earpod wattles' because pods are large and curl up (pictured left). Found in wetter forests, including tropical rainforests, especially in re-growth situations, and on the edge.

Scarlet Beans
Archidendron spp.

There are over 90 species of this genus found in tropical Asia, the south-west Pacific, and Australia. They grow as trees with compound leaves and many species are ramiflorous (flowering and fruiting on the branches). They produce spectacular sprays of flowers, in a range of soft colours, but usually white. They then develop large, curling, mostly very bright orange or red pods (pictured right, Cairns Botanical gardens) with contrasting black seeds.

Sea Beans, Matchbox Beans
Entada spp.

There are 30 species in this genus, and they are widespread throughout the tropics of the world. There are 2 species in Australia and they are both found in the tropical rainforests. ***Entada phaseoloides*** and ***Entada rheedii*** are some of the most obvious vines in lowland rainforest. The woody vines can become quite thick, and with age *Entada phaseoloides* develops a very distinctive series of sharp-edged twists (pictured below, Josephine Falls). The enormous pods (pictured right, Cape Tribulation) go from green to brown, then drop out, leaving an empty border. The large dark brown seeds are dispersed by the ocean and germinate all over the tropical Indo-Pacific. The attractiveness of the seeds and their waterproof quality made them an ideal container for matches.

Beans
Family Fabaceae/ Subfamily Papilionoideae

This is a massive family, with over 10,000 species in some 400 genera. They are found all over the world, and best developed in the tropics. There are over 1000 species in Australia, with 60 of those found in the tropical rainforest (Cooper and Cooper 2004).

Black bean *Castanospermum australe*. Only member of its genus. Leaves are bipinnate. Flowers appear in late dry season or early wet season, spectacular collection of orange and yellow that sprout from branches (pictured below, left). Fruits are huge woody pods, shaped like little canoes, break open to reveal 2 or 3 large brown seeds (pictured below, right). Seeds were eaten by Aboriginal people in times of lesser food availability, after much treatment. Pods similarity to canoes were used for exactly that by their children's dolls. Common in rainforests (and along watercourses in drier habitats) along east coast of Australia. Commonly planted as a street and park tree.

Mucuna spp.

This genus contains some 100 species spread throughout the tropics and subtropics of the world (Hyland *et al* 2002). There is 1 species found in the Australian tropical rainforest, **Burny Bean *Mucuna gigantea***. It grows as a climber, with compound leaves made up of 3 leaflets. The unusual flowers hang down and are greenish-white.

The following large pods also hang down, and are covered in irritating hairs The seeds are shiny, black and dented. They are famous amongst children in the region because if you rub the bean quickly on clothing, it gets hot enough to hurt your victim! Burny Bean is found in rainforest, especially along rivers, across tropical Australia, New Guinea, through tropical Asia into India and China (Cooper and Cooper 2004).

Order Fagales

Family Casuarinaceae

This family mainly contains the well known and widespread Casuarinas of Australasia. They usually grow as trees or large shrubs. Their foliage is made up of many drooping almost pine like 'leaves'. These in fact are not the leaves, but stems. The true leaves are virtually unnoticeable, being tiny and scale-like, and are found at regular intervals along the stems. The female and male flowers are also small and usually grow separately along the stems. They are followed by small, spiky, hard fruits. Like many plants in adjacent ecosystems, the casuarinas probably evolved from ancestors within the rainforest.

Gymnostoma spp.

This genus is believed to be the most ancient of the family and the ancestors of the more common Casuarinas. This genus had 18 species found on various islands of the South-west Pacific, and North-eastern Queensland, where there is one species, ***Gymnostoma australianum***. This species is only found in rainforest between the Daintree and Bloomfield Rivers.

Casuarina spp.

A genus of 17 species, found in South-east Asia, Australia and the south-west Pacific. They are named after the tropical rainforest bird, the Cassowary, because of the similarity of the tree's foliage to the bird's hair-like feathers. Only 1 species is commonly found associated with Australian tropical rainforest

Coastal She-oak, ***Casuarina equisetifolia***. Wispy tree with the thin foliage typical of family. Female flowers small and reddish, male flowers pale in colour and located at tips of stems. Followed by small, round, spiky, hard fruits. Most widespread species of Casuarina, being found along tropical coastlines of Australia, the Pacific and South-east Asia. Can be found on edge of rainforest where it grows near the coast.

Order Malpighiales

Family Clusiaceae

There are some 1,200 species in this family, and they are found throughout the tropics of the world (Cooper and Cooper 2004). The trees in this family found in Australian forests usually have very smooth, glossy leaves with fine venation that is almost perpendicular to the main central vein. They often have yellow latex when the leaves are torn.

Callophyllum spp.

Beauty Leaf *Callophyllum innophyllum*. One of dominant trees along tropical coastline of Pacific, including rainforest along coast of tropical Australia. Can grow quite large, the scaly trunk growing from coastal forest, then leaning over the sandy beach. Leaves are very smooth and glossy with fine, close venation. Flowers are white and yellow. Followed by round green fruits that end up on beach. Can be found all along wetter parts of tropical coast of Australia, and up into tropical Asia. Easily seen along beach at Cape Tribulation and at Cardwell.

Mangosteen
Garcinia spp.

Native Mangosteen *Garcinia warrenii*. Tree. Smooth leaves with fine venation, an intramarginal vein and yellow latex when torn. Usually noticed when drops its large, deep purple-mauve fruits. These fruits have calyx lobes and usually contain 2 or 3 large seeds in amongst sticky but edible reddish-purple flesh (pictured below). Seen in tropical rainforests of eastern Australia and up into New Guinea. Fruits are sometimes seen in wet season along the Cape Tribulation road and Mossman Gorge walking tracks.

Family Euphorbiaceae

This huge family has over 6000 species spread over 250 genera (Heywood *et al* 2007). They are found throughout the world, from tropical areas to temperate. On a global scale, the huge range of different species has few easily identifiable features. (It's status as a true 'monophyletic' - related species from a single origin family - group were in doubt, and recent research has confirmed these suspicions, taken many species out and placed them into their own families.) Locally however, many of the species tend to have small, dull flowers, rounded or heart-shaped leaves, and are commonly pioneer or edge species. There are over 150 species found in the Australian tropical rainforest (Cooper and Cooper 2004).

Aleurites spp.

There are between 3 and 6 species in this small genus (Hyland *et al* 2002). They are found from tropical Asia, to New Guinea and islands of the south west Pacific, to Australia (Cooper and Cooper 2004).

There are 2 species in Australia, and both are found in the rainforest. They are commonly called 'candlenut', as the seeds have a high oil content and were used as candles. ***Aleurites rockinghamensis*** is a commonly seen fruit on the forest floor (pictured right), especially in the mid altitudes, and they often have holes chewed in them by white tailed rats. ***Aleurites moluccanna*** is found from the Daintree and up into Cape York. Both species are found in New Guinea and south-west Pacific Islands (Cooper and Cooper 2004).

Bleeding Heart
Homalanthus spp.

There are over 20 species in the genus and they are found around tropical Asia, the Pacific Islands and Australia. 'Bleeding Hearts' are often noted for their large, heart-shaped leaves that turn a very bright red when they are about to be dropped. They also produce lots of small green fruits which turn a pretty purple colour when ripe. Bleeding hearts are a commonly seen plant as they colonise the edge of the forest and are initially fast growing. They can be seen along the Cape Tribulation road, and around the carpark at Lake Barrine. Their leaves provide food for the caterpillar of the largest moth in Australia, and a inhabitant of the tropcial rainforest, the Hercules Moth.

Cheese Tree, Buttonwood
Glochidion spp.

The several hundred species in this genus are found in Madagascar and the tropics of America, Asia and Australia (Hyland *et al* 2002). There are a dozen species in the Australian tropical rainforest (Cooper and Cooper 2004). They grow as shrubs and trees with simple alternate leaves. The flowers are usually small and white or yellow. They are followed by their distinctive feature; fruits that look like little pumpkins. Several species were used for their timber (Hyland *et al* 2002).

Breynia spp.

The 25 species in this genus are found in tropical Asia, Melanesia and northern Australia. The 2 to 4 species in Australia are found in various tropical forests. They are usually bushes with small rounded, alternate leaves. When crushed the leaves of some species, such as ***Breynia cernua*** and/or ***Breynia stipitata***, often give off a bad odour due to the presence of a methyl compound, and thus have the common name of 'fart bush'. Small red berries appear lined along the stems. They are common on the edge of the forest, such as walking tracks, and are common on certain sections of the Cape Tribulation road and the Bloomfield track.

Macaranga spp.

The Macaranga genus has about 280 species, found from Africa to tropical Australia. They are especially common and dominant in the rainforests of tropical Asia and the western Pacific Islands, where many of the heart-shaped leaved species have ecological relationships with ants.

There are half a dozen species found in Australia, and they are all found in association with tropical forests. One of the most commonly seen species, ***Macaranga tanarius*** has a centrally attached stalk and yellow venation. Being a 'pioneer' species it grows on the edge of the forest and thus is easily seen along the side of the road.

Family Passifloraceae

The family Passifloraceae includes over 600 species, found mainly through the tropics of the world. They grow as trees and shrubs, but of course are most famous as vines. In the Australian tropical rainforests, the leaves of species in this family are often heart-shaped or lobed, and often have two glands at the base. A winding tendril is used to climb, and this sprouts from a bud at the base of the alternate leaves, a distinctive feature of the family. The flowers are both striking and unique, due to the many frilled extensions of the perianth (the sepals and petals) that form the 'corona'. The fruits contain many seeds, which are surrounded by a pulpy aril (Heywood *et al* 1978).

Passionfruit vines
Passiflora spp.

There are over 370 species of passionfruit vines and they are found in tropical America, Asia and the Pacific, with some 8 native and naturalized species found in the Australian tropical rainforests (Cooper and Cooper 2004).

The flowers have the distinct frilled corona indicative of this family, 3 long styles that ride atop the central stalk, and 5 drooping stamens slightly below and from the same central stalk. These distinctive and elaborate flowers were named by Spanish missionaries for the passion of Christ; the 3 stigma represent the 3 nails, the 5 anthers the 5 wounds, and the frills of the corona represent the crown of thorns (Hutton and Cassio 1996). Many of the hundreds of passionfruit vines around the tropics produce edible fruits, but of these only a few have been cultivated, with South American species *Passiflora edulis* and *P. quadrangularis* providing the edible passionfruit. The former has several varieties and is found wild in the Australian tropical rainforest.

Adenia spp.

There are over 90 species in this genus, and they are found throughout the Old world tropics (Cooper and Cooper 2004).

Red Lacewing Vine *Adenia heterophylla*. (pictured right). Produces bright red hanging fruit that opens to reveal greyish-white angular seeds. Leaves provide food for caterpillar of the attractive Lacewing Butterfly. More common at lower altitudes, found in rainforest and monsoon forest throughout northern Australia, New Guinea and many Pacific Islands (Cooper and Cooper 2004).

Family Rhizophoraceae

While different mangrove species have evolved in several families, the family Rhizophoraceae is dominated by them. Several of these mangrove species are found on the edge of the rainforest, but that is the subject for another book in this series.

There is only one non-mangrove species of the family found in the Australian tropical rainforest.

Carrallia brachiata Like most of the other members of the family, tends to have opposite and round, smooth, shiny, leathery leaves with a short point, bunched together at the end of the branches. Also found in the adjacent woodland.

Family Salicaceae

Ryparosa spp.

There are 18 species in the genus *Ryparosa*, with only 1 occurring in Australia.

Ryparosa javanica. Usually small tree with long leaves, paler on underside. Most distinctive feature is it's cauliflory: the flowers are small, cream or green and borne in long inflorescences that hang off tree's bumpy trunk. Followed by hard, dark and velvety fruits. Tree said to be quite toxic, with flowers and fruits apparently poisonous. Usually only occurs in understorey of well-developed lowland tropical rainforest in tropical Asia and Australia. In Australia, is restricted to rainforest between Cape Tribulation and Daintree River, but can be locally common in that area, and easily seen in first hundred metres of Marrdja boardwalk.

Order Oxalidales

Family Connaraceae

Shell Vine, Conch Vine *Connarus conchocarpus*. Rarely noticed until it produces its distinctive fruit, which is small, red and shell-like. Fruit may be occasionally seen on forest floor, such as at Jindalba boardwalk, however usually difficult to tell which vine in the canopy overhead that it came from.

Family Elaeocarpaceae

The family Elaeocarpaceae has 300-400 different species around the tropics including the Americas, Asia, Pacific and Australia.

In the Australasian tropical rainforest, the species often have fine serrations around the leaf, swellings on both ends of the petiole at the attachments of the stem and leaf ('double pulvinus'), and many have tiny holes or hairs ('domatia') on the under surface of the leaf. The flowers are often white and many species have fringed petals, such as the flowers of *Elaeocarpus* (pictured below, Possum Valley) that result in a feathery or ragged appearance.

Quandongs
Elaeocarpus spp.

There are about 60 species of Quandongs and they are found in many parts of the tropics of the world, with about half being found in Australia. However, they are not at all related to the 'quandongs' of drier parts of southern and central Australia. They grow mainly as trees. There are often red leaves present both in the crown and on the ground beneath, and many species have fine serrations. The flowers usually have fringed petals and may hang down in large numbers. Most species produce attractive blue fruits that enclose a single largish seed.

Blue Quandong *Elaeocarpus angustifolius*. Most commonly seen and best known of Quandongs. Can grow into huge rainforest tree with well-developed buttress roots, but better known as slim, tall pioneer species on edge of rainforest. One of the fastest growing trees in the rainforest (Hyland *et al* 2002). Usually has several dropped red leaves underneath, and many marble-sized bright iridescent blue fruits (pictured right, Jindalba boardwalk). Underneath flesh is distinctive seed that looks like a brain. Tart flesh edible, but best when older and soft. Lightweight wood used for timber. Common on edge of rainforest along northern part of east coast of Australia.

Kuranda Quandong *Elaeocarpus bancroftii.* The fruits on this species are larger and duller than many other quandongs, and older fruits often have dull, flaky, cracked skin. Fruit has a persistent stalk and distinctive football shaped, pitted seed. Can be obvious plant along Dubuji boardwalk and Cape Tribulation road when fruiting.

Elaeocarpus culminicola. Leaves are slightly serrated. Profuse smelly flowers. Fruits on this species are small, commonly break off with twigs still attached, green to dull and brighter blue (pictured right, Cape Tribulation carpark). Distinctive seed is wrinkled with equatorial line and tiny gap at one end. Found at lower altitudes in north-eastern Queensland, through NT up to New Guinea

Order Rosales

Family Rosaceae

This family contains over 3,000 species and is found all over the world, with a centre of diversity in the northern temperate regions. Of course this group includes the rose, but it also includes some other very important species, such as apple, cherries, plums, peaches and strawberries. They are represented in the Australian tropical rainforest by 3 native and 1 introduced species of Raspberry Bramble. These tend to grow in clearings and on the edge of the forest as scramblers, and their thorny stems can make them annoying and painful to get through.

Rubus spp.

Rubus moluccanus. Large 3-lobed leaves and watery red colour raspberry like fruits (pictured right, Possum Valley). Found north-eastern Queensland, New Guinea, into South-east Asia. ***Rubus alceifolius***. Large many-lobed lobed leaf, with finely serrated edges. Raspberry like fruit. Introduced from south-east Asia, now in north-east Queensland. ***Rubus probus*** and ***Rubus queenslandicus*** are similar; both have compound leaves made up of smaller pointed leaflets, and have a strawberry like fruit. Former has rounder, shorter fruit and found in wet and dry forests along Queensland coast and up into New Guinea (Cooper and Cooper 2004). Latter has slightly longer fruit (pictured below, Possum Valley) and found only at higher altitudes of north-east Queensland.

Family Urticaceae

The family Urticaceae contains the stinging nettles of the northern hemisphere, as well as the most dangerous plants in the Australian rainforest, the dreaded stinging bushes and trees.

Stingers
Dendrocnide spp.

There are some 37 species of these plants and they are found in tropical Asia, the Pacific islands and Australia, with 6 species in the latter region, and they can all be found in tropical rainforest (Cooper and Cooper 2004).

The plants in this genus can grow as trees or shrubs. Most species in the Australian rainforests have large, heart-shaped leaves that are finely serrated and covered in tiny hairs. These hairs also extend along the reddish stems and cover the red fruits. The hairs are composed of glass-like silica, and also contain a poison. In the nastier species, the pain may re-occur for months. The plant produces a cluster of watery red or white fruits that resemble mulberry fruits and some are edible if you can avoid all the nasty hairs. Several species of insects live quite happily chewing on the leaves, as is usually evident with most specimens.

Stinging plants tend to be more common in light gaps in the forest and on the edge, such as roadsides and along walking tracks. Unfortunately, these are the very kinds of places where humans are liable to stumble into them.

Family Moraceae

The plants in the family Moraceae can include vines, shrubs and trees. They usually have alternate leaves and are characterized by the presence of latex, often white and copious, that can be seen when branches or leaves are broken. Another distinctive feature of the plants in this family is the presence of of two large protective leaf-like stipules at the end of the growing stem. Two tropical Asian species in the family supply large edible fruit and can be seen in various orchards in North Queensland; the 'breadfruit' and the closely related 'Jackfruit', the latter being the heaviest edible fruits in the world. The family has several useful plants, including the mulberry bushes for which the family is named. However, the most celebrated members of this family are the figs.

Figs
Ficus spp.

This genus contains about 1000 different species found worldwide, including as potted plants in many households. They are of one of the best known and widespread of all plant genera.

In fact, Figs are considered by some authors to be the most diverse genus of woody plants when looking at habit, growth forms and life forms (Kalko *et al* 1996). And because the genus is ubiquitous in all the world's rainforests, it is an often studied group, with some important and fascinating ecology. Many figs start off as mere epiphytes, with the 'strangler' forms of many species having their fruits deposited in the top of a canopy by a fig eating animal, and then grow their roots down around a host tree. Some species of figs in the Australian rainforests have evolve the habits of fruiting on the branches (ramiflory) and the main trunk (cauliflory).

Figs do not have flowers in the traditional sense; the tiny reproductive parts are hidden inside round receptacles that we normally recognize as the 'fruit'. Figs have a highly specific and reliant relationship with 'fig wasps' of the family Agaonidae: the tiny flowers of each particular species of fig are usually only pollinated by a particular species of 'fig wasp', while the wasps rely on the receptacle to reproduce inside (Compton *et al* 1996).

Because of this codependant relationship, it is thought figs have to provide these fruit /flower receptacles more or less continuously, and thus many figs fruit asynchronously. That is, trees fruit out of sync with other individuals of the same species (Kalko *et al* 1996). This also means that figs are being produced all year round, thus providing a regular food source for many animals in the rainforest; this is especially important when food is scarce at certain times of the year.

And thus figs are often considered 'keystone' species. They have traditionally been thought of as providing fruits mainly for generalist fruits eaters, and over 50 species of birds in Australia have been recorded feeding on fig fruits (Green 1993). However more recently it has been acknowledged that some figs may target specific dispersers and that there are some animals that may almost entirely feed on figs. It is for good reason that some Australian birds have the 'fig' prefix in their common name, such as the 'fig parrot' and the 'fig bird'; the latter bird is thought to be almost an obligate fig eater (Green 1993).

Plentiful Fig *Ficus copiosa*. Grows as a tree with sandpapery leaves. Produces attractive fruits; large and light green to red, covered in small bumps, sometimes with a little rough brown hair (pictured left). Fruits can be produced along the stem and trunk. Found from North Queensland to New Guinea, including other Pacific Islands.

Red-leaved fig *Ficus congesta*. Usually grows as a short tree. As common name suggests, red leaves in amongst usual green. Most distinctive feature is proliferation of fruits that grow off root-like structures low down on the tree, often at the very base (pictured right).

Hairy Fig *Ficus hispida*. Usually grows as medium sized tree that can develop buttress roots. Leaves often have a sandpapery feel. Plant gets its common name from fruit; small and round in shape, but slightly flattened, green or yellow with pale speckles and surface of sparse but stiff hair (pictured left), that looks and feels like stubble on unshaven face. Grows from the east coast of Australia through the Northern Territory, up into tropical Asia.

Climbing fig *Ficus pantoniana*. This unique fig actually grows as a vine, not a strangler (pictured right). Small, long leaves which when torn produce the distinctive copious latex of the genus. Found from North Queensland to New Guinea and the surounding Pacific Islands.

Banana Fig *Ficus pleurocarpa*. When seen as smaller tree on edge of forest or in revegetation, leaves have a slight single wave to edge, may have purplish colour underneath, and there are very long orange stipules at end of stems. Often grows as a strangler. Once a large tree deeper in the forest, is not noticed until large stipules and/or fruit have fallen onto forest floor (pictured below). Distinctive fruit, which can be coloured brown to red, but do indeed usually resemble squat, yellow bananas, with a green 'cap'. Usually found in well developed rainforest, in a range of altitudes, through most of the north-eastern Queensland.

Cluster Fig *Ficus racemosa*. Can be large tree that develops buttress roots. Deciduous. Common name comes from large 'clusters' of fruit that appear on main trunk and branches (pictured left). Ripen from green to yellow to red; not too bad to taste when soft. Found along watercourses in lowlands of tropical Queensland, and across northern Australia, and up into south-east Asia.

Septic Fig *Ficus septica*. Usually grows as small tree. Fruits are found growing on branches and trunk, sometimes low down. Fruits greenish at first, often ripening to a lighter green or yellow, are round with flattened top, slightly ribbed and covered in small white and brown dot bumps (pictured right). Also found in South-east Asia.

Variegated Fig *Ficus variegata*. Can be large tree that develops buttress roots. Fruits can be found along trunk or branches and are distinctive and attractive; start off green, ripen to red with green streaks (pictured left). Found from North Queensland, through the Pacific Islands, and into tropical Asia.

***Ficus virgata*.** Usually grows as strangler around larger host trees. Produces small round orange fruits that have distinctive nipple at one end (pictured right). Grows from North Queensland through to tropical Asia.

Family Rhamnaceae

There are some 925 species in this family are they are found throughout the temperate and tropical areas of the world. Many overseas species have spines, and this family includes the plant that is the 'crown-of-thorns' of Christian religion (Heywood *et al* 2007). There are some 160 species in Australia, and 20 of those occur in the tropical rainforest (Cooper and Cooper 20045). The leaves are simple and are usually alternate. The undersides have obvious venation and often have pale or silvery undersides due to minute hairs (below, left). In the Australian rainforest species, the flowers are usually tiny and coloured yellow and/or white (pictured below, right). The stamens mainly arise from just outside the central disc, usually opposite and in between each of the pointed petals.

Alphitonia spp.

There are over 20 species in this genus and they are found throughout the tropics of Asia, the Pacific and Australia, including 5 species in the tropical rainforest of Australia (Cooper and Cooper 2004). They grow as trees, not always large, and several species favour disturbance and are thus common on the edge of the forest. As typical for the family, the leaves are pale underneath. If the leaves are crushed in several species, they have a toothpaste like smell, and are known as "Sarsaparilla Trees'. The most commonly encountered species of the lowlands, ***Alphitonia excelsa*** is also known as a '**soap tree**', as it has enough saponin in the leaves so that when they are crushed and rubbed in freshwater, they produce froth. The flowers are tiny, and usually shades of yellow and white (both pictures above are ***Alphitonia petriei***, which has relatively narrow leaves, and is common on the edge of the rainforest at higher altitudes, Longlands Gap). The following fruits are small brittle capsules that attract various fruit pigeons.

(Eurosids II)

Order Malvales

Family Malvaceae

There are over 1000 species in this family and they are found all around the world. The most well known species in this family are the cotton bush and the garden 'Hibiscus'. Members of this family in Australia are obvious and common in more open habitats, such as the tropical woodlands; some of the smaller species occur on the edge of the tropical rainforest.

Cottonwood or Beach Hibiscus, *Hibiscus tiliaceus* (pictured above, right). Normally grows as a small tree with heart-shaped leaves. Produces large flower that usually only lasts for a day, goes red, then drops. Another type of plant in the same family, ***Thespesia*** (pictured above, left), is very similar, occurring in same beachside habitat, has heart-shaped leaves and a yellow flower; however, the leaves on Hibiscus are paler underneath and flowers have protruding stamens from the top of the central column, while *Thespesia* lacks this. Common along roads, rivers and beaches in tropical lowlands. Where they occur along larger rivers such as the Daintree and Russell/Mulgrave, the flowers are often seen floating on surface of water.

Family Sterculiaceae

This large family has some 1000 species found all over the world. The seeds are often contained in dry fruits such as pods. The family contains the plants that provide us with chocolate and the original cola beans.

Sterculia spp.

This genus has between 150 and 300 species and they are found throughout the tropics of the world (Hyland *et al* 2002). The genus name *Sterculia* is named after the Roman god of dung heaps (!) due to the smell of the flowers of certain species (Cooper and Cooper 2004).

The 3 species found in the tropical rainforest of Australia also occur in the monsoon forest and in forest along waterways in the drier woodlands across tropical Australia, and 2 species also occur in New Guinea. The main species seen in Australian rainforest is the **Peanut Tree *Sterculia quadrifida***. It produces large orange-red fruit that opens up canoe shaped sections to reveal black seeds (pictured right, Cairns Botanical Gardens). These seeds are edible, quite delicious, and taste rather like peanuts; thus the common name.

Kurrajongs
Brachychiton spp.

This genus has over 30 species, most of which are found in Australia. Kurrajongs are more common and better known as open woodland trees, where they are also known as 'bottle trees'. However, there are 11 species found in the tropical rainforest (Cooper and Cooper 2004). The leaves are often heavily lobed, and many of the rainforest species have red flowers. They develop woody pods that split open in canoe-shaped sections to reveal they are packed full of seeds.

Native Flame Tree *Brachychiton acerifolius*. Tree with bulbous trunk that loses its leaves as it produces a spectacular show of bright red flowers. Sometimes seen in hills of coastal areas as patch of red contrasting against green forest. Easily seen as street tree, especially common around Mossman (pictured left).

Tulip Oaks
Argydendron spp.

One of the more common rainforest trees are the 'Tulip Oaks', which are of course not at all related to the northern hemisphere Oaks. They have a distinctive 'rocket ship' shape to the trunk, with flanged buttress roots and a straight, tall trunk. The roots and trunk have a unique reddish shade (pictured below, left). Every few 'mast' years there is an explosion of their distinctive seeds; small round seeds with a single large wing (pictured below, right) that produces a whirled flight when released from the top of the tree. Many do not whirl that far, and the trees are often surrounded by a small carpet of seedlings.

Commersonia spp.

Brown Kurrajong *Commersonia batramia*. One of most common and obvious trees growing along side of roads through lower altitude rainforest. Flowers in summer, looking like creamy coloured 'snow' on branches. Develops small brown spiky fruit. This tree's habit on the side of the road, its flowering makes and the pale underside to the leaves make it look very similar to the unrelated *Alphitonia*. However, this latter plant has much paler undersides to the leaf, and small, smooth (not hairy) and brittle hard fruits.

Order Sapindales

Family Sapindaceae

There are over 2000 species in this family, with most found in the tropics and subtropics of the world. There is often a small terminal leaflet that has been reduced to a spine, and the junction of the leaflet stalk where it joins the main leaf stalk is often swollen ('pulvinus'). The most famous members of this family are tropical Asian fruits such as Lychee and the Rambutan. In the Australian tropical rainforests, the family is represented by over 100 species. They are mainly trees and shrubs with compound leaves. The fruits split open to reveal seeds which are often mostly covered in a brightly coloured fleshy aril. It includes many different genera that are referred to as 'Native Tamarinds'..

Harpullia spp.

There are over 25 species in this genus, and they are found from tropical Asia, through the Pacific to Australia (Hyland *et al* 2002). The genus name is Indian in origin (Cooper and Cooper 2004). There are 8 species in Australia, and 6 can be found in the tropical rainforests (Cooper and Cooper 2004).

Dwarf Harpullia *Harpullia frutescens*. Small shrub, with distinctive flattened out or 'winged' stems. Produces small white and pink flowers. Followed by orange or red fruit that split open to reveal two black seeds covered in orange aril in two compartments. Usually found in understorey of relatively undisturbed tropical rainforest of north-eastern Queensland.

Diploglottis spp.

There are 10 species in this genus and they are centred in the south-west Pacific. There are 10 species in Australia, and 8 can be found in the tropical rainforests (Cooper and Cooper 2004). Most species have green to yellow fruits that split open in three to reveal 3 large seeds that are often covered in an orange or red fleshy aril.

Smith's Tamarind *Diploglottis smitti*. Grows as a tree, often develops buttress roots. Mainly found in the better developed tropical rainforest in the lowlands in north-eastern Queensland (Cooper and Cooper 2004), such as around Mossman Gorge

Family Meliaceae

This family has some 600 species found throughout the tropics of the world, with only a few in temperate areas (Cooper and Cooper 2004). It is well represented in the Australian tropical rainforests with almost 40 species, and they include some of the more impressive trees in the forests here. Some species grow simple leaves, but many have compound leaves, commonly divided pinnately or binnately, with the leaflets being opposite and having terminal leaflets. The fruits often split open to reveal seeds with a colourful aril. The cut wood of many species has an odour and the family includes some important timber trees, including the true African Mahoganies, as well as various local regional 'mahagonies'. The most famous local timber is the 'Red Cedar', which was pursued by the timber getters all the way up the east coast.

'Mahogany'
Dysoxylum spp.

There are about 80 species in this genus, and they are found from tropical Asia to Australia (Hyland *et al* 2002). There are some 15 species found in the Australian tropical rainforests (Cooper and Cooper 2004). Most species here have fruits that spilt two to five ways to reveal seeds with a colourful aril.

Spur Mahogany *Dysoxylum pettigrewianum*. Can grow to very large tree. Trunk has many small bumps (lenticels). Develops many fruits along its branches that are at first dark with light spots. They ripen to orange and open up to reveal red and white red seeds (pictured right, Daintree Hideaway). One of more common larger trees of coastal lowlands of North Queensland, and are also found in New Guinea and other Pacific Islands.

Yellow Mahogany *Dysoxylum parasiticum*. Can grow as tall tree, often with buttress roots. Cauliflorous: develops masses of flowers on trunk, followed by orange fruits (pictured left). These open up to reveal foam white inside with red, white and dark brown seeds (pictured below). Found at a range of altitudes in north-east Queensland, Cape York and up in New Guinea and tropical Asia. Good example with sign on walking track at Hypipamee Crater.

Family Rutaceae

The family Rutaceae includes a wide range of trees, shrubs and vines. Members of the family are found all over the world, with many temperate and tropical species. The plants have leaves that are simple or compound, but if the latter, the leaflets are opposite. The compound leaflets are often in groups of 3, ('tri-foliolate'); sometimes the groups of leaflets are reduced to just one, such as in many *Acronychia,* and are effectively simple leaves. The most distinguishing consistent characteristic of this family are the translucent dots within the leaves that can be seen if held up to the light. If the leaves are crushed, the oil dots produce a citrus smell. This is the family that supplies the citrus fruits such as oranges, mandarins and lemons.

'(Australian) Maples'
Flindersia spp.

There are about 17 species of *Flindersia*, and they are only found in New Guinea, Australia, New Caledonia and associated islands (Cooper and Cooper 2004). Of these, there are a dozen species found in the rainforests of Australia. Although often called 'Maples', these trees are not at all related to the northern hemisphere maples. They are often huge trees that are usually only noticed when they drop their distinctive fruit. They open up their usually 5 arms, (looking like a woody, knobbly octopus), to let go of their many light, papery winged seeds (upper photograph). However, it is seems even more common to see the unopened fruits on the forest floor (looking like giant spiky gherkins) as storms or cockatoos often knock them from the canopy (larger picture below). The genus is named after Mathew Flinders the explorer. Many species were used for their timber.

Acronychia spp.

This genus contains some 50 species, found mainly in the tropical forests of Asia, Pacific and Australasia. They are usually trees, often with buttress roots. The 'leaves' are usually opposite and when crushed have the citrus-like odour of the family. The scientific name refers to a feature of the flowers; a closer look at some will reveal a re-curved tip of the petals. Many species have edible fruit, and are also popular with forest birds such as fruit pigeons, but they taste very tart like the lemons of the same family. The fruits are usually light purple, orange, white or, most commonly, yellow. They often have part of the stem still attached (below). There are at least a dozen species in the Australian tropical rainforest.

Lemon Aspen *Acronychia acidula*. One of the most commonly encountered species in the genus. Found from inland of Cape Tribulation down to Eungella, but most common around mid and higher altitude rainforests, such as regrowth forest around Lake Eacham.

Evodia
Melicope spp.

There are some 230 species in this genus, and they are found from Madagascar, through tropical Asia to the Pacific (Hyland *et al* 2002). The leaves are usually made of three leaflets, and these sets are opposite each other. The fruits are usually hard, small pitted or dimpled capsules that split open to reveal small dark seeds. There are about 10 species in the Australian tropical rainforest.

Pink Evodia *Melicope elleryana*. Attractive plant that usually grows as bush or small tree. Often develops soft and crumbly corky bark and has semen-like smell when cut. Grows tri-foliolate leaves and beautiful pink flowers that erupt along branches that attract butterflies and birds. Followed by small, hard capsules with black seeds. Commonly planted on road sides and gardens for both its pretty flowers and leaves that are the food for the spectacular bright blue Ulysses Butterfly.

Family Anarcardiaceae

This family has about 600 species found throughout the tropics of the world, with 8 species being found in and around the tropical rainforests of Australia (Cooper and Cooper 2004). Species in this family often have clear or milky sap that may turn black when exposed to air. There may be irritants present in this exudate which can cause severe skin reactions in some people. The sap can be found in different parts of the plant depending on the species, including the bark, twigs, leaves, and even parts of the fruit. Despite this, one of it's species is adored in the tropics for it supplies the different varieties of the delicious Mango. The family also includes the pistachio and the cashew (Heywood *et al* 2007)

Mango *Mangifera indica* (pictured right). Originally from tropical Asia, has been planted in and around the suburbs, roads and farms of tropical Australia. Was planted not just for its fruit, but also for timber, shade, and showy display of flowers. Has now become naturalized in rainforest and woodlands in much of tropical Australia.

Blepharocarya spp.

This genus is endemic to Australia, and has only 2 species (Cooper and Cooper 2004). The genus name means "eyelid nut" (Cooper and Cooper 2004), referring to the eyelash like bract (pictured below).

Rose Butternut *Blepharocarya involucrigera*. Compound leaves. Produces strange but distinctive woody bracts (pictured below, Lake Barrine) which then release small fruits that resemble seeds surrounded in hairs. Found in rainforest, monsoon forest and rocky woodland from Torres Strait down to Atherton Tablelands (Cooper and Cooper 2004)

(Asterids)
Order Ericales

Family Lecythidaceae

Most of the at least 450 species in this family are large trees in the tropics, with a centre of diversity in the Amazonian rainforests (de Wit 1967). The species in this family often have large and spectacular flowers made impressive by a spray of many stamens, such as *Barringtonia*. Many fruit into hard capsules that contain seeds, as in the most famous member of this family, the 'Brazil Nut'.

Barringtonia spp.

This is a genus of about 40 different plants found throughout the tropical Indo-Pacific region. The plants usually have large simple leaves that are clustered at the end of the branches. The flowers are often spectacular: large sprays of stamens, sometimes hanging down in racemes. These spectacular displays of flowers are thought to be pollinated at night by animals such as bats and moths. The following fruits vary; in the Australia tropical rainforest, they can be either fleshy or, if dispersed by fresh or salt water, large and woody.

Mango Pine, Cassowary Pine *Barringtonia calyptrata.* Can grow to large tree. Large rounded leaves in whorl at end of branches that are typical of genus. Produce long hanging branches of cream flowers (right) followed by large blue fruits with persistent calyx 'tail' (below). Found in lower altitude forests of North Queensland, through Cape York, New Guinea.

Beach Barringtonia *Barringtonia asiatica.* Usually short, but sometimes wide spreading, tree. Large rounded leaves in whorl at end of branches that are typical of genus. Produce large upright flowers, many long cream stamens with pink ends (pictured left). Flowers drop during day. Followed by very large woody, angular fruits that wash up on beaches. Found in rainforest when along seashores throughout tropical Indo-Pacific.

Family Sapotaceae

The family Sapotaceae has at least 40 genera and 600 species (de Wit 1966). They usually grow as trees and are found mainly in the tropics (de Wit 1966).

One of the main features of this family is, as suggested by the name, the presence of sap. They produce (often large) fleshy fruits with a large single seed or several small seeds. These seeds sometimes have a scar on the side. The family includes the edible fruits 'star-apple' and the 'sapodilla' and the plant whose latex was used to make chewing gum (de Wit 1966).

Plum Boxwood, Brown Pearwood *Niemeyera prunifera* (pictured above). Produces large blackish fruit with light red flesh and one large single seed. Seed has distinctive 'cut' diamond shape across side. Found in rainforest of eastern Queensland (Cooper and Cooper 1994).

(Euasterids I)

Family Boraginaceae

There are some 150 genera and 2,500 species in this family (Cooper and Cooper 2004). The leaves are often covered in small stiff hairs that may give it a rough feel. The flowers usually develop in umbels.

Cordia spp.

There are 250-300 species in this genus (Rohwer 2002), named after a German professor from 500 years ago (Cooper and Cooper 2004). They often have heart shaped leaves, and as they are sometimes rough to the touch they may be known as 'rough leaf' or 'sandpaper plants' (Rohwer 2002). The flowers grow in umbels, with most species producing small, white flowers, although the best known species, the ornamental Scarlet Cordia, *Cordia sebestena*, has large colourful red flowers. The small round fruits often develop from a persistent green calyx at the base. There are 5 species found in tropical Australian forests (Cooper and Cooper 2004); some have fruits that are sticky when squashed, thus 'glue berries'.

Trumpet tree *Cordia subcordata*. Grows as shrub or small tree. Large, heart shaped leaves slightly rough to touch. Flowers attractive; orange, with six crinkled petals. Fruit is woody, round, brown, few centimetres across, develops in 3 pointed persistent calyx, (unlike other local species in genus, fruits are woody reflecting sea dispersal). Inside is harder shell with several tiny edible seeds. Found in tropical forests when on edge of seashore, especially on islands, from tropical Indo-Pacific.

Order Gentianales

Family Gentianaceae/Loganaceae/Strychnaceae

Strychnos spp.

Of the 170 species of *Strychnos* found around the tropics and subtropics of the world, 3 species are found in Australia's tropical rainforests (Cooper and Cooper 2004).

Strychnine vine *Strychnos minor.* Grows as a vine. Usually only noticed when small orange fruits found on rainforest floor. Being 'strychnine', they are of course not edible.

Fagraea spp.

The genus Fagraea includes small trees and climbers, including 2 species that are common in the lowland rainforests of north east Queensland. They usually have tubular flowers and produce fleshy fruits.

Porcelain Fruit *Fagraea cambagei*. Normally grows as small, thin tree. Tubular flowers followed by one of the prettiest fruits of rainforest floor; this 'Porcelain fruit' is a smooth, shiny round ball coloured a delicate pink or white. Common along Marrdja boardwalk.

***Fagraea berteroana*.** Usually grows on other trees as an epiphyte, or even on rocks. Produces tubular flowers that are either white or yellow. Followed by orange-red oblong fruits that have a strange cellophane-like peeling. There is an example growing on main epiphyte laden tree in Mossman Gorge car-park.

Family Apocynaceae/Asclepiadaceae

This family has over 4,500 species (Cooper and Cooper 2004) and is well represented in the tropics of the world. The most distinctive feature of these species is the presence of milky white latex in broken branches and leaves. The family includes such tropical ornamentals as the 'Allamanda vine', with its pretty yellow flowers, which can be seen in many tropical gardens. This family is well represented in the Australian tropical rainforest, with 30 genera and over 100 species (Cooper and Cooper 2004).

Chain Fruit
Alyxia spp.

Alyxia usually grow as shrubs with whorls of small leaves that are slightly prickly to the touch (and are thus sometimes called 'prickly alyxia'). They develop small red or orange fruit that are often joined together in short 'chains' or in connected lumps. They are found in rainforest, monsoon forest, coastal heath, vine thicket and woodlands in eastern and tropical Australia and Asia. They are very common in the rainforest understorey along the tracks at Mossman Gorge.

Gondola Bush, Banana Bush
Tabernaemontana spp. (formerly *Ervatamia*)

There are over 100 species in this genus and they are found throughout the old world tropics, from Africa to Australia (Cooper and Cooper 2004).

They usually grow as short shrubs with longish, opposite leaves. As members of the family Apocynaceae, snapping off a leaf will produce some white latex. The flowers are white and jasmine-like, with twisted petals. They produce very distinctive fruit that are usually yellow or orange in colour, and shaped like a small gondola or banana, although they are not at all related to bananas. They can be found in tropical open woodlands and forest, and there are 2 species found in Australian tropical rainforests, with ***Tabernaemontana pandacaqui***, (pictured left) being the most common one in tropical rainforest from NSW to New Guinea.

Milky Pine or Milkwood *Alstonia* spp.

There are over 40 species in this genus, and they are found from Africa, through Asia to Australia (Cooper and Cooper 2004). 'Milkwoods' get their common name from the milky latex that appears when the bark on the trunk is cut into; this is poisonous and may cause blindness (Brock 1997). Despite the common name, they are not pines, but often grow straight and tall like one. They have small white flowers and then develop long, thin seed pods with fluffy material around the seeds. In Australia, there are about half a dozen species, and they are found in the tropical forests and savanna woodlands (Cooper and Cooper 2004).

Cerbera spp.

This Old World genus contains 7 species that are found from Madagascar, across tropical Asia and into Australasia. Many species produce colourful and fleshy, but toxic fruits. There are 3 species that can be found in Australian tropical forests.

Cassowary Plum, *Cerbera floribunda* and **Grey Milkwood *Cerbera inflata***. Can grow as tall trees and exude latex when damaged. Large, long leaves arranged in erect whorls, many coloured red. Fruit are large, bright blue and egg-shaped, with *C. floribunda* (pictured) being larger and brighter. Both these fruits are typical food for cassowary, resulting in common name for *C. floribunda*. Latter species occurs in North Queensland, up into Cape York and into New Guinea, while *C. inflata* occurs only from Atherton Tablelands down to Innisfail (Cooper and Cooper 2004). Cassowary Plum is common at Mossman Gorge and along Jindalba boardwalk.

Dog Bane, Rubber tree *Cerbera manghas*. Tree. Produces beautiful large flowers that are white with a red centre and have strong pleasant scent. Followed by large egg-shaped red fruits. Usually grows in rainforest right on beach from tropical Australia up into Asia. Can be seen around Cape Tribulation beach.

Order Rubiales

Family Rubiaceae

This is one of the largest of all flowering plant families with over 7000 species. They are found mostly in the tropics of the world, but are also represented in the subtropical and temperate regions. The leaves are usually either opposite to each other, or arranged in whorls. They have a small line on the stem that runs between the base of the two opposite leaves, called the 'interpetiolar scar' (Jackes 2001). The flowers in this family are varied, but many species are quite pretty, often white and frequently fragrant. The ovary in the flower is inferior and because of this, many of the resulting fruits have persistent calyxs that appear as lobes, tubes or rings at one end of the fruit. A family of this size is bound to have important economic applications: because of the attractive flowers and their scent, many species are used as ornamentals, but this family provides a product that this author feels proves the wonder and majesty of nature: the beans for coffee. As in the other tropical rainforest areas of the world, the Rubiaceae is well represented in the Australian tropical rainforest.

Myrmecodia spp.

One of the most unusual members of the Rubiaceae family are the 'ant plants'. There are about 26 species, and they are found in New Guinea, western Pacific Islands and tropical Australia, with 3 species found in the latter region (Cooper and Cooper 2004). They are so named for housing ant colonies within their main swollen stem. They are found on trees in mangroves, paperbark wetlands, and tropical rainforest. These funny looking, round blobs can be seen in the canopy along Marrdja boardwalk and in the Cairns Botanical Gardens.

Nauclea spp.

Of the 10 species in this genus found around tropical Asia and the western Pacific, only 1 speices is found in Australia (Cooper and Cooper 2004).

Leichhardt Pine *Nauclea orientalis*. A flowering plant not a conifer. Can grow to be large tree. Leaves large and dark glossy green, prominent yellow venation, edges frequently curl; often dropped in the drier season. Flowers small with orange-yellow petals, protruding white stamens; stick out collectively from central point, giving inflorescence the look of small spiky balls (pictured right). Slowly produces squishy yellow blob that is actually collection of small fruits that have followed the many small flowers; about size of golf ball, edible. Found in forests, woodlands of eastern, northern Australia, through to tropical Asia.

'Gardenia'

Gardenias are one of the best known plants in the family Rubiaceae. The common name 'Gardenia' is applied to several different genera that have had their species shifted around in recent years, thus there are several plants in the Australian rainforest that are referred to as 'Gardenias'. They usually have large, white, and fragrant flowers.

Gardenia actinocarpa. Small shrub, with longish thin leaves. Usually only noticed when shrub produces its showy flowers. Like many other gardenias, flowers are white and fragrant, with petals being longish and thin, often curling up at end. Separate male and female plants. When sparser female plants do produce the brownish to greenish fruit, they have longitudinal ridges (thus the specific name *actinocarpa*, which means 'star-shaped fruit'), and long persistent calyx lobes. Endemic to Daintree region, found only in tropical rainforest in the lower catchments of the Noah Creek and Oliver Creek area; can be seen along Marrdja boardwalk.

Attractocarpus spp.

Brown Gardenia *Attractocarpus fitzalanii*. Usually small tree, with smooth, longish rounded-end leaves that have oil dots that can be seen with lens (Cooper and Cooper 1994). Flowers are fragrant and with overlapping white petals. Followed by large, smooth round fruits which develop on tree over long period, ripen slowly from green to brownish to yellowish. Flesh is quite thick and hard, and there are many seeds set in pulp inside; this pulp is slightly edible, but tastes quite terrible. Found in rainforest, usually near water, and on edge of mangroves, along tropical eastern Queensland coast.

Hairy Gardenia *Attractocarpus hirta*. Small shrub with hairy leaves that give plant its common name. Most distinctive feature are beautiful large white flowers which stand out in gloomy understorey of rainforest. Flower (lower picture) is made up of five petals that overlap at base and has yellow centre; tips are pointed when fresh, but become rounded and curl up with age. Following fruit is small, orange, hairy and with distinctively long calyx (upper picture). Endemic to understorey of tropical rainforest in north-eastern Queensland. Can be seen along Marrdja boardwalk.

Order Lamiales

Family Lamiaceae

October Glory *Faradaya splendida*. Vine. Large, opposite leaves and showy white flowers that do indeed appear around October. Following fruits also large; look, smell and feel like potatoes (which are not related and not even a type of fruit). If a piece of flesh is picked at, fruit darkens. Cassowaries eat the fruit (Hinton and Hinton 1989), and is edible for humans. Used by local Aboriginals as 'indicator' plant; when it fruits, it is the time the Brush Turkeys have eggs in their nest (Breeden and Cooper 1993). Found in canopy and on edge of rainforest in north-east Australia and New Guinea.

Clerodendrum spp.

These plants are usually seen as shrubs or small trees. Species in this genus can be found in the adjacent woodlands as well as the rainforest. The local species have white flowers with very long, protruding, almost string-like stamens. They produce round fruits presented on a bright red flower like bract.

Family Verbenaceae

This family has undergone recent revision and lost many of its species to the related family Lamiaceae.

Lantana spp.

Lantana camara. Usually grows as scrambling, dense bush. Stems are often hairy, leaves are serrated and give out scent when crushed. Flowers are attractive bunches of yellow, pink and red, followed by small dark purple berries. Can be seen in many disturbed areas throughout wetter tropics of Australia. Despite attractive flowers and attraction for butterflies and honeyeaters, this is one of worst weeds in Australia..

Family Oleaceae

This family contains over 550 species around the world. They grow as trees, shrubs, climbers and herbs, usually with opposite leaves. They often have white fragrant flowers, such as jasmine. They may develop fleshy blue or black fruits, such as olives. Within these large bunches, the fruits often develop in pairs. The seeds often have some longitudinal lines.

Native Olive ***Chionanthus ramiflora***. Tree. Longish dusty blue fruit that distinctively wrinkles, may have persistent stalk. Found in woodland and forest, eastern Queensland, New Guinea, Asia (Cooper and Cooper 2004).

Family Bignoniaceae

This family contains many tropical trees and vines, often with tubular flowers. The family includes ornamental species such as the striking purple flowered Jacaranda, seen as a street tree in the higher altitude towns of the Atherton Tablelands, such as Atherton.

African Tulip Oak ***Spathodea campanulata***. Large tree. Originally from tropical Africa but now planted throughout tropics. Most noted for prolific production of large and beautiful cup-shaped orange/red flowers that can appear for much of warmer times of year. Common in parks, gardens and along roadsides throughout tropical Australia.

Golden Bouquet Tree ***Deplanchia tetraphylla*** (pictured right). Small tree with huge rounded leaves in a whorl. Large bunches of yellow flowers. Found in various forest types in North Queensland and up through Cape York to New Guinea, seemingly more common in coastal situations, and also planted in gardens.

Jungle Vine ***Neosepica jucunda***. Common vine usually fairly high up in canopy. Only noticed when drops its small tubular purple flowers (pictured below left). Woody fruit split open to reveal lots of light, flat, winged wind-dispersed seeds (pictured below, right).

Order Solanales

Family Solanaceae

This family contains over 2600 species in 90 different genera. Many of the plants have spines and spikes on the leaves and stems. The flowers have five sepals and five petals, which are often fused (Perry 1977). It is a very important family for humans, providing many of our foodstuffs, including tomato, capsicum, chilli, paprika, eggplant, and one of the most important food staples of them all, the potato. These plants also provide other products, such as nicotine for tobacco and other alkaloids. These compounds are also what render many species in this family quite toxic and the fruits and leaves have been known to poison children and animals. Because of this, the family has often had a negative reputation, and is sometimes dubbed the 'deadly nightshade family', with many species given such adjectives as 'devil' and 'deadly'. The family has a centre of diversity in the tropical rainforest of the Americas, but it is found worldwide and include many aggressive weed species, especially in disturbed areas.

'Wild Tobaccos', 'Devil', 'Kangaroo Apples', 'Wild Tomatoes', etc.
Solanum spp.

This is one of the largest of all genera of plants, with over 1500 species (Perry 1977). They are mainly herbs, shrubs and small trees. They often have furry or spiky leaves and the crushed leaves are odorous, reminiscent of the related tobacco. The flowers are usually purple or white, with crinkled-looking petals and sepals (Grieg 2000). The petals are sometimes pointed, and the flower often has a bunch of yellow stamens in the middle. The flowers are followed by the fruits which are normally small, round, shiny, multi-seeded and are usually toxic. The genus is used all over the world for various medicinal practices, both ancient and modern. Some species of 'Kangaroo Apples' have been grown for the extraction of the alkaloid 'solasodine', the starting point for the production of the making of steroidal hormones (Lassak and McCarthy 1997), which are involved in the production of contraceptive pills (Low 1997). This genus also provides many different types of 'eggplant', but undoubtedly the most important contribution of this genus to the world is the Neotropical species *Solanum tuberosum*, the 'potato'. There are many Solanum found in the tropical rainforests of Australia, including both native and weed species. Most are 'weed'-like in habit, being small and usually found on the edge or in disturbed areas.

Family Convolvulaceae

This family includes many vines, such as the 'Morning Glory'. It also provides humans with the tropical Asian water plant, the sweet potato.

Morning Glory
Ipomea spp.

This is a large genus with many introduced and native species. The leaves are usually heart shaped. Flowers can be white, red or orange, but the best known are blues and purples. They are normally deep tubular flowers with large spreading petals joined together to form the familiar 'morning glory' flower. As a vine, they grow in the rainforest canopy, but are usually only noticed by humans on the edge of the forest.

Cyclone vine, ***Merrima peltata***. This distinctive vine has large heart shaped leaves. Tends to take over in areas where there are frequent disturbances, such as storms and cyclones. Most common on rainforest edge on small hills exposed to coast in lowlands. Obvious south of Cairns on Bruce highway, and along sections of Cape Tribulation road. Also occurs in tropical Pacific.

(Euasterids II)

Order Apiales

Family Araliaceae

This family includes over 1450 species of large trees, smaller shrubs, or vines. They are found mainly in the tropical forests of Asia, Australia and the Americas, but are also well represented in temperate areas, by, for example, some of it's most famous members, such as Ivy and Ginseng *Panax* spp. In the Australian tropical rainforests the species often have compound leaves, and are distinguished by the swelling of the stem where it clasps to the main branch. They often have their flowers and fruits arranged in umbels.

Scheflerra spp.

This genus has some 100 species throughout the tropics of the world, with 3 species found in the tropical rainforests of Australia.

Umbrella tree, Octopus Plant, *Schefflera actinophylla*. Best known in the genus. Usually grows as climber on other trees, roots resembling a strangler fig. Once established, may grow as large tree in it's own right. Called 'umbrella' tree due to circular arrangement of leaflets. Flowers held in umbels along long 'octopus' arms held at end of main stalk, high above other branches and leaves. Small hard red fruits develop along same arms. Found in forests of tropical Australia and New Guinea. Specimens can easily be seen along edge of rainforest, and around Lakes Eacham and Barrine.

Polyscias spp.

There are 100 species in this genus and they are found throughout the old world tropics, from Africa to Australia, with 10 species found in the latter region (Cooper and Cooper 2004). This genus includes several plants known as 'celerywood' and 'basswood'. They are usually tall, spindly trees, with leaves made of a series of large opposite leaflets that make a distinctive silhouette (pictured below). Some species have the fresh smell of celery when their branches are broken, thus the first common name. They are often common pioneer species and so can be seen along the side of roads through the rain-forest.

Family Myodocarpaceae

This family has just recently been erected and split from the Araliaceae, based on both molecular studies and morphological features of the fruit (Heywood *et al* 2007). There is only 2 genera and 19 species in the family, and they are centred in New Caledonia, except for a 2 species that are also found in New Guinea, other islands in the south-west Pacific and north-eastern Queensland (Heywood *et al* 2007).

Delarbrea spp.

This genus has 6 species, with 1 found in Australia (Cooper and Cooper 2004).

Blue Nun *Delarbrea micheana.* Usually grows as a small, rather spindly plant. Compound leaves, with leaflets increasing in size towards end of stem. Noticed when it produces huge collection of stunning bright blue fruits. Fruits favoured by Satin Bowerbirds for decoration (Cooper and Cooper 2004).

Family Pittosporaceae

This is a family of 9 genera and about 200 different species found in Africa, Asia, Australia and the Pacific (Hyland *et al* 1999). Some of the genera are found only or mainly in Australia. The plants in this family can be trees, shrubs or climbers. They have simple leaves that are usually alternate or sometimes appear whorled (Clarke and Lee 1987). The flowers are often tubular and with five sepals and five petals (Clarke and Lee 1987); many are pale coloured and emit pleasant odours at night (Perry 1977). Because of their pleasant flowers, some species in this family, especially Pittosporums, are used as ornamentals.

Pittosporum spp.

This is a genus of between 150 to 200 different species found in Africa, Asia, Australia and the Pacific, with about 15 found in Australia (Hyland *et al* 1999). The name is usually pronounced as a quick 'pit-toss-spore-um'. In the Australian rainforests they are usually shrubs or small trees with large rounded leaves. The flowers are usually pale and at the end of the branch. The following yellow or orange fruit have points at the end and split open to reveal red seeds.

Hairy Pittosporum *Pittosporum rubigosum*. Usually shrub of only few metres high of rainforest understorey. Specific epithet and common name refers to rusty hair on underside of leaves (Nicholson and Nicholson 1994), which gives them a soft feel. Collection of white tubular flowers at top of plant. These develop into slightly hairy yellow/orange fruits with a ball of red sticky seeds inside.

Order Asterales

Daisies
Family Asteraceae

This is one of the biggest of all flowering plant families, with somewhere between 13,000 to 25,000 species around the world, including over a 1000 species found in Australia (Cooper and Cooper 2004). They are easily recognized by their flowers; the actual 'flower' head is composed of many different tiny flowers. These flowers (called 'florets') are usually arranged in two whorls, the inner whorl made up of tiny flowers, the outer whorl with one large petal-like sepal (Clarke and Lee 1997). In many species the two lots of whorls of flowers may be of different colours. In some species, the outer whorl or the inner whorl may be absent (Grieg 1999). Most species are the small herbs known as 'daisies', but the family also includes other well known plants such as thistles, everlastings and Chrysanthemum. Other species provide food for humans such as artichokes, endive, lettuce, and of course sunflower seeds. It is one of the most widespread of all flowering plant families, with species found from the tropical lowlands to the coldest mountain tops. They are also widespread in Australia, where they can be common in open habitats such as grasslands and woodlands. As daisies are primarily herbaceous and such plants are rare in the lower light understorey of undisturbed tropical rainforest, natives are not dominant in undisturbed rainforest. However, there are 30 to 40 species associated with tropical rainforest (Cooper and Cooper 2004), including many weeds that have established themselves on the edge and disturbed parts of the forest.

Singapore Daisy *Sphagneticola trilobata*. Small herb that grows to few feet in height. Often forms thick spreading mats. Leaves serrated, sometimes subtly lobed (thus the specific epithet *trilobata)*. Usually only noticed when produces bright yellow daisy flower at top. Originally from South/Central America. Grows on edge of various forest types along east coast of Australia, from Cape York down to NSW (Cooper and Cooper 2004) Major weed in wet tropics, often found in stands at edge of rainforest, along roads and in carparks.

(Various unclassified)

Family Balanophoraceae

How this strange little family is related to the other flowering plant families is still unresolved. There are some 45 species and they are found throughout the tropics of the world, with concentrations in the Asian and American tropics. The bulk of the plant is underground and in some species can grow as 'large as a babies head' (Heywood *et al* 2007). They are leafless plants that live as parasites, most species attach to the roots of trees. A phallic structure erupts from inside the underground plant and then bursts through the ground to flower and then fruit. This is the part of the plant that humans observe, usually at the base of host plant near their roots. Most people assume the round ping-pong ball sized and shaped structure is a mushroom, and the scientific name means 'acorn bearing' because of the shape (Cooper and Cooper 2004). Most species develop very tiny flowers, amongst the smallest flowers of all plants. Little is known about their reproductive strategies, but several overseas species are known to develop eliasomes on their seeds for dispersal by ants (Heywood *et al* 2007).

Balanophora spp.

This is the main genus in the family, and it has 15 species, including 1 found in the Australian tropical rainforest, ***Balanophora fungosa*** (pictured below, Jindalba boardwalk), that also occurs throughout the tropical Pacific and South-east Asia to India (Cooper and Cooper 2004). These bizarre plants are probably best seen at Mossman Gorge. The flowering part of the plant appears around late winter, and the small mushroom like ball has a ring of male flowers around the base, and tiny female flowers on the 'head'.

Worms, Molluscs, Arachnids Crustaceans & Insects

Earthworms and Leeches
Phylum Annelida

The name 'worm' is a vague term used for many different types of animals, including major groups such as the flatworms, and for the ribbonworms, as well as being used for the larvae of various insects that look vaguely worm like. However, the word worm is probably best applied to the Annelids. They lack any appendages or antennae (Harvey and Yen 1995), and are known as the 'segmented worms', for a close look reveals the body is divided into many tiny segments. The annelids include 3 large classes; the marine polychaete worms, the earthworms, and the leeches. The latter 2 classes have species found in the Australian tropical rainforest.

Earthworms
Class Oligochaeta

Earthworms are found worldwide in most terrestrial ecosystems, but are more abundant in wetter habitats. They are well known for their importance in turning over soil as they burrow and feed on detritus and organic material. They are more commonly seen on the surface of the soil after a lot of rain, as they are avoiding drowning by the rising water table (Cannon 2000). At most other times they can be seen in any digging of dirt or broken log in the rainforest. The most spectacular species in north-east Queensland is the 'Giant Blue Earth worm', and is exactly that; giant (about 2 feet long) and blue. Like other worms, it only comes out after very heavy rain and is more common at higher altitude.

Leeches
Class Hirudinea

Leeches are distinguished from earthworms by the suckers at the ends of their body and the way this feature results in their inchworm like movement. The larger posterior sucker enables them to attach to the substrate, and their prey.

Leeches are specialized worms that have evolved to either eat prey or parasitize other animals, by sucking blood out of them. The latter have a series of small jaws that cut into the skin of the host. The digestive juices of certain species may anaesthetize the skin so they can feed for some time without being noticed, and have anticoagulant properties to let the blood flow (Harvey and Yen 1995). There are different types of leeches that attack different animals; fish, frogs, birds and mammals, including humans hiking through the rainforest.

Phylum Mollusca

This is a very diverse group that include a range of different looking animals; chitons, nudibranches, cephalopods (squid and octopus), bivalves (clams, mussels, oysters) and gastropods (snails, seashells). Molluscs are essentially marine animals, but over time some species have evolved into the intertidal zone, estuarine waters, up into freshwater streams, and finally onto land (Gooderham and Tsyrlim 2002). Species in the latter 2 groups may be seen around tropical rainforest, although they tend to be quite cryptic.

Class Gastropoda

This group includes most of the seashells and all of the snails. As the animal grows, it adds layers to the shell, resulting in the familiar spiral pattern. The soft animal inside includes a single enlarged 'foot' and a head that sticks out of the shell. The head usually has 2 eye stalks and 2 sensory tentacles (Gooderham and Tsyrlim 2002). Gastropods have an organ called a radula, which has evolved into different tools for different species; in the herbivorous snails, it serves as a grinding tool to scrape off plant material.

While the marine snails (subclass Prosobranchia) have internal gills and can seal up their shell with an operculum, most of the land based snails found in Australian rainforest belong to the subclass Pulmonata and have internal lungs and no operculum (Harvey and Yen 1995). Due to the warm and moist conditions, the tropical rainforests of north-eastern Queensland have the highest diversity of land snails on the continent, with over 300 species, including many with very restricted distributions (Stanisic 2000).

Class Bivalve

Bivalves are soft-bodied animals that live within 2 main shells attached on one side to allow opening and closing. This group includes the oysters, clams and mussels. They are most diverse and abundant in the cooler oceans of the world, but there are a few species in the fresh waterways of the Australian tropical rainforest. Species of any size belong to the family Hyriidae.

They look quite similar to marine mussels, however, they are rarely seen as they burrow into the finer sediments of sand and mud under the water. While the foot anchors most of the shell in the sediment, the hind edge of the shell remains exposed and the siphons filter feed the passing water for nutrients. They initially start their life as tiny parasites on the lungs of native freshwater fish (Gooderham and Tsyrlim 2002).

Phylum Chelicerata

This phylum includes the spiders, scorpions, mites and ticks, and they are all placed in the one class.

Class Arachnida

Most species in this group have two main body parts, the cephalothorax and the abdomen. The first part contains the head, including the eyes and mouth, and the legs; most species in this group have four pairs of legs as adults. The second section, the abdomen, contains the reproductive and digestive organs.

Scorpions
Order Scorpionida

Scorpions are easily recognized by the long tail with a sting at the end that is held erect, and the huge pedipalps of which the final segments have evolved into very robust crab like pincers called 'chela'. They are not commonly noticed in the rainforest, but they do have the unusual habit of fluorescing under ultraviolet or black light. The best place to see them is at the Kuranda Venom Zoo (pictured right)

Harvestmen
Order Opilionida

Harvestmen are very similar looking to spiders. They tend to have quite small round bodies; unlike spiders, their body is a single unit rather than having a wasit between the cephalothorax and abdomen. Like spiders, they have 8 legs, the second pair of legs often being much even longer than the others, as well as 2 pedipalps. The legs are usually very long and whip like, and almost appear too delicate. They usually only have only 2 pairs of eyes and these are often raised above the body as two bumps.

Unlike spiders, they cannot produce silk and lack spinnerets. They were known as 'harvestmen' because in the northern hemisphere they came out during harvest time; however in the tropics they can be seen at any time. They catch their own invertebrate prey, but also scavenge.

Spiders
Order Araneae

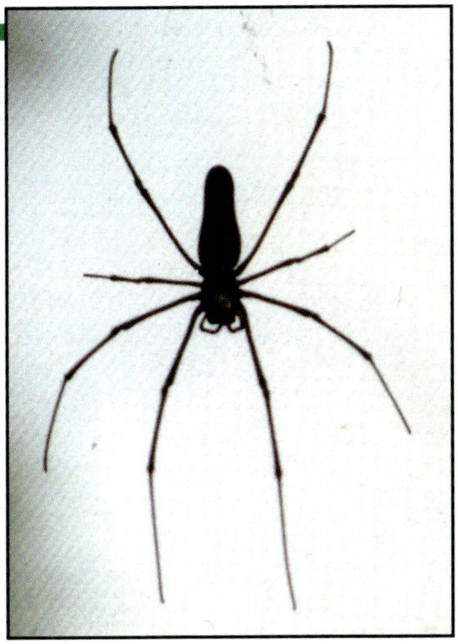

Spiders have 8 legs as well as an extra pair of appendages at the front of the body called the pedipalps. In females these are mainly used in manipulating food, but in males they are bigger and used to transfer sperm to the female. In general, female spiders are bigger than their male counterparts; in certain species, this can be an extreme size difference. Thus many males have to be very careful not be eaten by the female, with some presenting packages of food to keep the female busy while they mate.

The production of silk by spiders is well known and no doubt one of the most important evolutionary developments for this group. Two spinnerets are located on the back of a spiders abdomen and this is where they release their silk. There have been 7 distinct types of silk identified, with different silk used by different species for various purposes and stages of their lives. Of course, some of the more advanced groups use it to construct elaborate webs to snare their prey. Most spiders are venomous and catch their prey by subduing them by injecting venom. This is delivered through fangs at the ends of another pair of stout appendages of the front of the head, known as the Chelicera. Some species are infamous for having venom which is potent enough to kill a human being. Fortunately, most species found in the tropical rainforest are harmless to humans. Australia appears to have one of the highest diversities of spiders in the world, and North Queensland is recognized as having the highest diversity of spiders on the continent (Raven 2000). The spider order is divided into 2 main sub-orders. The more primitive is the Orthognatha, (the 'mygalamorphs'), and the other sub-order is the Labidognatha.

Mygalomorphs
Sub-order Orthognatha

The Mygalomorphs are mainly ground dwellers, often burrowers, and include some of the largest and heaviest of all spiders, including the pet tarantulas of the Americas, and in the Australian rainforest; the whistling and 'bird-eating 'spiders, and the funnel-web spiders.

This suborder is distinguished externally by the arrangement of the chelicera, for they need to lift their head up dramatically to strike down with their large weapons (the chelicera of the more advanced sub-order can pivot sideways, providing a wider biting span).

Sub-order Labidognatha

This is the major suborder of spiders. While the mygalomorphs just have simple 'book lungs' for respiration, this group has also further evolved a tracheal system within the body (Preston-Mafham and Preston-Mafham 1993).

Huntsman Spiders
Family Sparassidae

This large widespread family includes some 1000 species, with about a tenth of that number found in Australia (Brunet 1996). In terms of span, they include some of the largest spiders in Australia. They are flattened in appearance, with long, pointed legs that bend forward, and have efficient grip. This enables them to scurry sideways very fast, cling onto walls and ceilings, and hide under objects.

These appearances and behaviours are generally an adaptation to their life underneath the bark of trees. The fact that they do not use webs, and that they may chase down their prey, gives them their most common name. However, most probably use a 'sit and wait' hunting strategy, placing themselves on tree trunks and preying on the passing traffic. The Huntsman family can potentially be found all around the world, but are there are more species in the tropics. Huntsmen are often seen on the walls inside lodges, toilets and other outside buildings in the Australian tropical rainforest. Their ability to hide under objects, their often large size, and their alarming speed, means they can be initially frightening for the first time visitor. And probably the second time visitor, too.

Lichen Huntsmen
Pandercetes spp.

This is a small but distinctive genus of spiders in the Huntsman family. They can be found in the tropical rainforests of South-east Asia, New Guinea and Australia. In this habitat, they are found tight against the lichen covered bark of rainforest trees.

They are rarely noticed, for the hairs on the body, the brown and green colouration, and the pressing against the trunk to reduce their shadow, all helps them camouflage against the bark (pictured below, Cairns Botanical Gardens). They tend to sit still, but if you get too close, they may scuttle quickly around to the other side of the tree trunk. Some species have egg sacs that also look much like lichen on the tree, which the adult then guards (Clyne 1977). There are some large species in the tropical forests of Asia. A species in the Australian rainforest is called ***Pandercetes gracilis***.

Wolf Spiders
Family Lycosidae

Wolf Spiders are mainly ground dwelling spiders coloured in natural shades of greys and browns. They have the usual 8 eyes, including 2 larger ones. Different species are found in a range of habitats all around the world, but they seem to be more obvious in open areas. This is likely due to the fact that they are open range hunters, chasing down their prey on the ground.

Many shelter in burrows by day and hunt by night. Their eyeshine can easily be seen by torchlight in clearing or roads on the edge of the rainforest. This is being reflected by the 2 larger eyes, due to their eyes being sensitive to polarized light (Brunet 1997), an attribute that may help them to orientate themselves in the open by the moon and starlight even on cloudy nights. In the tropics of Australia, 'wolf spiders' have been reported biting, killing (and possibly consuming) cane toads (Raven 2000).

Water Spiders or Nursery Web Spiders
Family Pisauridae

This family contains over 25 species in 10 genera (Brunet 1997). They are often large spiders that are generally brownish, with some species in the genus *Dolomedes* having orange or red markings on the body. They look similar to their relatives the Wolf Spiders. The main difference is the eyes; while Wolf spiders have two large eyes and a row of smaller four eyes arranged in a square, water spiders have two rows of smaller eyes. As suggested by the name, they are found in or near water over much of Australia. They are capable of running across the surface of the water or diving under. In this habitat they feed on fish, frogs and aquatic invertebrates. They are also referred to as 'nursery web' spiders because the female keeps her eggs in a slik sac held close to her body, and when it is almost time for them to hatch, she builds a small 'nursery web' between reeds or other waterside vegetation and aggressively guards them.

The Giant Water Spider *Megadolomedes* spp. (pictured left) are probably the most impressive of the family, for they can grow to the size of a human hand They are usually seen on the edge of freshwater streams with their head down and their front legs in the flowing water. They hunt for fish and aquatic invertebrates and are capable of hunting on the surface of the water. They are found all over Australia, and can be common along freshwater streams through lowland tropical rainforest, where they may be noticed on bridges and causeways across water.

Family Araneidae

There are over 2,500 species in this family (Preston-Mafham and Preston Mafham 1998) and they are found worldwide in almost every terrestrial habitat. The family includes many species of garden spiders, and thus they are probably the most familiar of all spiders. They build intricate webs, often in pleasing symmetrical patterns. And even though this family includes the biggest of all spiders in Australia (by span), they are normally admired rather than feared, perhaps because they stay in their web 'in plain sight', rather than scuttling around.

St. Andrews Cross spiders
Argiope spp.

The 25 Australian species in the genus are amongst the best known and admired spiders in the bush. Many species are also common on the edge or gaps of the rainforest. They are often colourful, with stripes, and can increase the brightness even more by contracting the abdomen and raising the hair on their body (Brunet 1996).

They are probably best recognized because they often sit with each pair of legs together, giving the impression of a four legged spider. Their pleasing webs will often have an obvious 'stabilmentum', a criss-cross pattern of much thicker web. One species, ***Argiope keyserlingi***, builds four such structures in the shape of a cross in the web, thus giving rise to the common name for the genus. This structure was originally presumed to be for added strength or to blend the spider into the web, but experiments proved it was the opposite; the obvious structure exists to alert birds to not fly into the web, saving the spiders their resource and the bird from entanglement (Preston-Mafham and Preston-Mafham 1993).

Spiny backed spiders, Jewel Spiders
Gastercantha spp.

These are some of the prettiest web inhabiting spiders in Australia. They are very small with short legs and a tiny head, but with a broadly expanded and hardened abdomen. This is usually colourful with bright patterns and bizarre shapes, including elaborate and long spines. They are common on the edge and in gaps of the rainforest.

Gasteracantha fornicata, (pictured left). First spider scientifically described in Australia, by the crew of Cook's ship *Endeavour* (Brunet 1993)

Golden Orb Web spiders
Nephila spp.

The Golden Orb Web Spiders are among the largest of the spiders to build aerial webs. They sit in the middle of their large 'orb' or circular shaped web usually with their legs spread out and head pointing downwards. The abdomen is elongated and often decorated with blacks, yellows and whites. These spiders are referred to as a 'golden orb web' not because of the colours sometimes found on the spider (that can include golden yellows), but because, at the right angle, the yellow silk of the web takes on a very delicate golden colour. Most specimens of Golden Orb Web spiders observed are the females. This is because the *Nephila* genus has the biggest size difference between females and males of any spiders; the females can be anywhere between 100 to nearly 1000 times bigger in weight than her partners (Preston-Mafham and Preston-Mafham 1998). In the photograph below the small male is on top of the female. As wel, there are often other types of spiders in the web which are commonly mistaken for males; the 'dew drop' or 'quicksilver' spiders of the *Argyodes* genus scavenge off the smaller prey caught in the web that the giant *Nephila* doesn't bother with. This symbiotic 'kleptobiotic' relationship results in the larger spider having a cleaner web (Brunet 1998). Golden Orb Web Spiders are very common in the lowland tropical rainforest, especially after the summer rain. There are several species in this genus across eastern and tropical Australia.

Nephila pilipes (previously ***Nephila maculata***) (pictured below) is best known and largest. Longish, silver coloured abdomen, yellow or gold colour on the cephalothorax, bright red on the pedipalps, and long thin, black legs, sometimes with some orange bands. There are a couple of smaller (but still quite large!) species: ***N. Plumipes*** and ***N. edulis***, that often have more colourful yellow abdomen, white on the cephalothorax, and more colourful black and orange/yellow banded legs.

Hygropoda spp.

Hygropoda spp. are slim spiders with striped bodies and long legs with small but obvious stuff hairs near the joints. ***Hygropoda lineata*** is a local rainforest species that may be seen on the surface of large green leaves (pictured right).

Jumping Spiders
Family Salticidae

This is the largest of spider families, with over 4000 species (Preston-Mafham and Preston-Mafham 1993), and about a tenth of those are found in Australia (Raven 2000). They are found worldwide but it is in the tropics where they really reach their highest diversity. In the rainforests they among the prettiest of all spiders, with a range of colours and patterns (photographs below). As suggested by the common name, jumping spiders leap about with much energy, and precision. They have two very large eyes at the front and top of the head and some species are reported to have sight acuity second only to the birds of prey (Raven 2000). Their head and abdomen are quite separately moveable, with the head often turning at an angle to the abdomen, and swivelling to look directly up at their human observers, a rather disconcerting but endearing behaviour from such a tiny animal.

Species seen in the Australian tropical rainforest include the striking copper coloured and blue striped ***Cosmophasis bitaeniatus*** (lower photograph on right) and the often bright **Green Jumping Spiders *Mopsus* spp** (larger photograph on left)

Phylum Crustacea

Crustaceans are, like spiders and insects, arthropods. This means they are invertebrates with a hardened outer casing and jointed appendages. They range in size from planktonic to enormous rock lobsters. There are over 50,000 species throughout the world (Jones and Morgan 2000), but despite the huge range of size, shapes and forms, crustaceans have certain features in common. Three main body parts can usually be recognized in most crustaceans; the head, the thorax and the abdomen, with the first two sections often being referred to as the cephalothorax. A key feature of the crustaceans is the extent to which they have evolved many different appendages of their head and body for many different specialized functions. Generally, crustaceans have two pairs of sensory antennae at some stage of their life, (as opposed to insects that have one pair), and the head usually also has usually three pairs of appendages for food manipulation and feeding. Other appendages off the thorax can be for a range of other functions, including walking, swimming, fighting or even holding eggs.

Crustaceans are most dominant in the oceans and seashore, and are not well represented in the tropical rainforests. In this habitat, they include slaters, freshwater crabs, prawns and crayfish. The freshwater lakes, rivers and streams of North Queensland, much of which goes through rainforest, have the highest diversity of freshwater crustacean species in Australia (Davie and Short 2000). The slim River Prawns *Macrobrachium*, spp. can be fairly abundant in the lowland streams, the biggest being the **Giant Jungle Prawn *Macrobrachium lar*** (pictured above), which is found in permanent flowing streams in high rainfall areas throughout the Indo-Pacific (Davie and Short 2000). Several species of the more familiar Yabbies *Cherax* spp. are found throughout the rainforest streams, while the Spiny Crayfish *Euastacus* spp., are generally restricted to the cooler waters of the higher altitudes, where there is high species endemism.

Phylum Uniramia

Millipedes

Class Diplipoda

Being long animals with many legs, millipedes are superficially similar to centipedes. They of course do not have a thousand legs, but usually have more legs than centipedes, often several hundred. They have 4 legs per body segment, 2 on either side, and the segments are more rounded than centipedes. Unlike the carnivorous centipedes, most species feed on vegetable matter. Because of their less challenging diet, they are not aggressive like centipedes, are generally slower moving, do not have poison glands, and if disturbed, tend to roll into a ball or play dead. There are about 15 orders around the world, with at least 9 of these found in Australia (Harvey and Yen 1995).

Centipedes

Class Chiliopoda

Centipedes are long, slightly flattened, many segmented arthropods with lots of legs. Despite the name, most species have far less than 100 legs, but they certainly have many, with 2 per segment, one of either side. The legs can twist fast and the animal can scuttle along quickly. Centipedes tend to have longer thinner antennae than millipedes. The 2 legs of the last segment have been modified into longer gripping or sensory claws, while the 2 legs of the first segment have been modified into a pair of poisonous claws ('forcipules') just behind the head (Ruppert and Barnes 1994). They use these weapons to subdue their prey, which mainly includes invertebrates, but sometimes also small frogs, reptiles and mice (Harvey and Yen 1995). Thus, centipedes have should be treated with respect, for several species have been implicated in delivering nasty bites to humans. The are 5 distinct orders are all found in Australia, including the very fast and long legged Scutigerid 'house' centipedes, sometimes seen in caves or around rocks in the rainforest, and the widespread leaf litter dwelling Sclopendrids (photograph below).

Class Insecta

The name 'bug' is often used to describe insects (and sometimes other creepy-crawlies), but the word 'bug' should really only be used for one group of insects, the sapsuckers of the order Hemiptera. This is only one order in a class of two dozen other orders that comprise one of the most diverse groups of organisms in the world. They range in size and shape from giant stick insects to tiny beetles (pictured below). More than a million species have been identified and there are many more to be discovered. In Australia, over 85,000 species have so far been identified, and it is thought that there is at least an equal amount yet to be named.

Insects are arthropods and therefore have a hard exterior skeleton (the exoskeleton), body segmentation, and multiple appendages. They are usually differentiated from other arthropods by their three pairs of legs, one pair of antennae, usually three body segments, and the possession of wings. The evolution of wings and flight is generally thought to be a high contributing factor to the incredible success of the insects. With such huge diversity, there are exceptions to these basic body plans, or examples where these features are not so obvious.

The highest diversities of insects are usually associated with the tropical rainforests of the world. It is estimated that 90% of all the worlds insects are found in the tropical rainforests (Osborne 2000), and in Australia it is no different. The wet tropics region of North eastern Queensland is usually recognized as having the most diverse insect fauna on the continent. The following sections describe the orders that are commonly encountered in the Australian tropical rainforest.

Dragonflies and Damselflies
Order Odonata

This order is considered one of the oldest of all insect groups. A now extinct type of dragonfly was the largest and heaviest insect to have ever flown, with a wingspan of over 70cm (Silsby 2001). The Dragonflies and Damselflies are both long-bodied insects with two sets of long, delicate-veined wings, and they are often beautifully coloured. They tend to perch on low vegetation near water, where they occasionally fly off to predate on another insect (pictured below, right). Male dragonflies and damselflies have two small projections at the end of their abdomens called claspers, while females tend to have a swelling at their end. When they mate the use their claspers to grab each other in an unusual 'wheel' position. Most species have aquatic, predatory larvae. Despite the sometimes slightly fearsome appearance of the adults and the hideous appearance of the juveniles, they are completely incapable of harming humans.

Dragonflies and damselflies are placed into different suborders, and there are some main differences to look out for in the adults. Pictured below is a typical dragonfly (upper larger photograph) and a typical damselfly (smaller lower photograph). Dragonflies are generally larger, stouter bodied insects, while damselflies are usually more slender and elegant. The shape of the wings varies; in dragonflies the wings are uniformly broad, but in damselflies the wings taper towards the body. In dragonflies the two pairs of wings may be different in shape and size, while in damselflies they are usually similar. In posture, dragonflies generally rest with their wings outspread, while Damselflies generally rest with their wings held back along their body; but of course there are always exceptions (see next page).

Damselflies
Suborder Zygoptera

There are over 100 species of damselflies found in Australia, and they are organized into a dozen different families.

Family Diphlebiidae

This small tropical Asian/Australasian family includes the Rock Masters ***Diphlebia* spp.** Unlike most damselflies they tend to rest with their wings outstretched. They are quite large, relatively thick bodied for damselflies, and the males can be bright blue. The thin transparent black wings often have black patches, and thus may give the appearance of 'paddles'. They are usually seen perched on rocks in fast fresh streams, such as at the Mossman River.

Family Coenagrionidae

This large, worldwide family of damselflies are often referred to as the 'Pond Damsels'. They are thin, and the males have quite striking bright blues and reds with contrasting black. They are frequently seen on the side of still bodies of water, landing on waterside or floating vegetation (pictured left, Daintree Eco Lodge).

Dragonflies
Suborder Ansioptera

There are over 200 species of dragonflies found in Australia. They are classified into 18 families (Theischinger and Hawking 2006).

Family Libellulidae

This is the largest family of dragonflies. Most species are dominated by warm colours such as yellows, oranges, reds and browns. They can be common around pools of water, including in disturbed areas, such as clearings and water-filled ditches along roads.

Cockroaches
Order Blattodea

In contrast to the introduced cockroaches, many native cockroaches are not quite so disgusting, and in fact many are quite attractive. Cockroaches are usually fairly subtle creatures, coloured in browns to avoid detection. They are dorso-ventrally flattened, and the part of an insect normally at the neck called the pronotum is in cockroaches enlarged to a broad sheild protecting part of the body and the head. The legs are spined and adapted for fast scurrying (Zborowski and Storey 1998). Unlike the more omnivorous house cockroaches, many native species mostly likely feed on dead vegetation, such as rotting logs, fallen leaves, and other detritus in the rainforest.

Giant burrowing Cockroach, *Macropanesthia rhinoceros*. The heaviest cockroach in the world (pictured below). Wingless. Female feeds young on dead leaves dragged down into underground tunnels where it generally lives (Zborowski and Storey 1998). Adults may emerge after heavy rain. Sold as a pet. Found in wetter forests, north-east Queensland.

Termites
Order Isoptera

While termites are often called 'white ants' and do superficially resemble ants, they are in fact not at all closely related. Their order is considered much more primitive than ants, but they have evolved a complex colonial 'caste' system similar in some respects to ants, with specialized 'worker' and 'soldier' (lower photograph) individuals. Termites feed primarily on cellulose in dead vegetation and are thus acknowledged to be important recyclers in the rainforest.

Termites are quite inoffensive creatures to larger animals like humans. They are quite delicate and are rarely seen in the rainforest, most living within the soil or vegetation. Relatively few species construct mounds, but the species that do are usually the most noticed. In the rainforest small dark mounds may be noted in the branches of the canopy, and on the forest floor there are small round mounds (pictured right) inhabited by species of *Termes.* These ground mounds are common along the Daintree boardwalks and around the grounds of Kingfisher Park, and are used by nesting Paradise Kingfishers.

Praying Mantids
Order Mantodea

Praying Mantids are vaguely stick insect-like in appearance, but are distinguished by the modified front limbs, which can spring out to grasp and manipulate their prey. When folded up, they give the appearance of 'preying' and thus the common name. They are efficient predators, catching and munching on anything they think they can tackle, including other mantids. For most species, males usually mate several times in a lifetime (Rentz 1996); occasionally the female mantid may bite the head off the male during sex, but he can keep going, as the nerves that control his movements are located in the abdomen (Horne and Crawford 2005). After mating, the female produces an unusual foam-like egg case that hardens into a stiff papery cocoon called an ootheca, and this can be found attached to vegetation or rocks.

Giant Mantids
Hierodula spp.

The most distinctive feature about Giant Mantids is their size; they can grow up to 8.5cm, although smaller species and instars can be seen around tropical gardens. They are usually green or brown and unlike the garden mantids, their 'neck' is usually flanged. They will often have pink or orange colour to the inside of their limbs, especially the front ones. They can be seen hunting in the lowland rainforests by day and night, and they have even been known to catch and eat small frogs.

Crickets, Grasshoppers and Katydids
Order Orthoptera

Like many common names cricket and grasshopper mean different things in different places. Generally, the crickets tend to be nocturnal, have long antennae and females have an extension to the back of their body which is stuck into soil to lay eggs (ovipositor). In contrast, grasshoppers tend to be diurnal, have short antennae and females lack the ovipositor. Generally, the nocturnal crickets are duller coloured, while the diurnal grasshoppers can be brighter colours; however, in the tropical rainforest crickets are often colourful bright greens (pictured).

The insects of this order are well known for their quick springing and jumping abilities, a skill possible due to the incredibly long legs that are bent back and to the side of the body. These legs are also involved in sound production; they are rubbed to create the sometimes very loud noises that males make to attract females. This is also a group of ecological and economic importance, as they constitute some of the most dominant grazers in many ecosystems and are among the more serious crop pests around the world, such as some locusts.

King and Forest crickets
Family Stenopelmatidae

This is generally considered the most primitive family and also includes the largest crickets of them of all, the 'Wettas' of New Zealand. They are usually seen at night at wetter times of the year.

The Giant Forest Crickets, *Papuaistis, Penvalva* are some of the most striking of Crickets. Some species have white patches where the legs bend, and are thus commonly known as 'white-kneed crickets' (pictured right, Bartle Frere). They are large brown crickets with very long antennae and when disturbed hop about in every direction, usually landing on the person who is most scared of them.

Katydids
Superfamily Tettigonioidea

Katydids are also known as 'long-horned grasshoppers', but are classified with the crickets. They are common in tropical rainforests worldwide, but because they are often high in the canopy, and many species are green and frequently shaped like the leaves and stems in amongst which they live, they can be hard to detect. Only their loud calls give away their presence. The more commonly seen species are those that live in the grassy edges of the forest, such as the Snout-nosed Katydids. They are so named because of the snout that projects forward past the eyes (pictured). Their calls are the loud strident buzzing that can be heard on summer nights from the grassy edges in the lowland rainforests.

Stick insects
Order Phasmidae

This order contains some of the most amazing of all insects. The scientific name 'Phasmid' refers to ghost and alludes to their blending in with the surroundings so well that they are virtually invisible to the visitor. And as the common name suggests, these insects often grow to resemble stems, sticks and leaves, with some of the tropical forms growing to enormous sizes. The impersonation of vegetation doesn't stop there; the females often drop eggs that look like little seeds. In most species, the females are usually larger and wingless, and the males are smaller, winged, and fly about in search of females to mate with. Because of their cryptic habits, adults are usually only noticed in situations when they are seen away from their natural habitat such as climbing about on a picnic table.

They are an unusual group of insects for two other reasons. One is 'autotomy'; that is, many species can regenerate lost limbs, something which is relatively unusual among insects. Another unusual feature of this group is the high incidence of pathenogenesis; that is, populations of certain species that have been found to reproduce without males.

There are some 150 named species of phasmids in Australia (Hangay and German 2000) and as with many groups, the highest diversity is probably found in the mix of rainforests and other ecosystems in North-east Queensland. There are 2 families in Australia; the family Phylliidae (the difficult to see and rare 'true' leaf insects, as well as some stick-like leaf insects) and the much larger and dominant family, the Phasmatidae.

Family Phasmatidae

Subfamily Tropidoderinae

Spiny Leaf Insect or **Macleays Spectre Stick Insect *Extastoma tiaratum***. Not a true leaf insect as such, but a stick insect which happens to look like a leaf. Very first instar mimics an ant. As they grow, the younger stages look very much like a leaf, with a tail that curls up over the body (pictured). As adult, females are thick-bodied and can grow up to 15 centimetres long. Popular pet overseas, but in wild are endemic to the forests and woodlands along east coast of Australia.

Subfamily Phasmatinae

Titan Stick Insect *Acrophylla titan*. Longest of all Australian phasmids, indeed longest of all Australian insects, and one of longest insects in the world; can grow up to 25 centimetres long. Coloured brown and looks just like a stem, thus blends in well with it's environment, not normally noticed until found in another context, (such as pictured right). Seems to be most common in forests and woodlands along coast of Eastern Australia. Another similar species ***Eurycnema goliath***, is a more tropical, greener, slightly thicker version of this species.

Subfamily Platycraninae

This family includes several species in the genus *Megacrania* found only on *Pandanus tectarius* plants on some tropical Pacific Islands. One species lives in Australia.

Peppermint Stick *Megacrania batesii* . One of most interesting and iconic insects in Australian tropical rainforest (below). Can often be detected by chewed leaves of Pandanus plants on which they feed, and the large amount of insect poo ('frass') they leave. Slender animals that can be different shades of green, to almost blue. Common name derives from the smell of peppermint when disturbed; don't lean in too close to smell them, or may squirt you in eye. Believe me. Found only on *Pandanus* plants in coastal swamp and forest around Cape Tribulation and Etty Bay near Innisfail.

Bugs
Order Hemiptera

The name 'bug' is often used in common usage to talk about many different 'creepy crawlies'. However the name 'bug' actually refers to a specific order of insects. Bugs tend to be rounded in shape with hardened forewings that often fit back over the body rather like a shield. They differ from the superficially similar beetles by their usually shorter antennae and the presence of a pointed mouthpart. Most species use this to suck sap from plants, or predate on other insects.

Cicadas
Family Cicadidae

The biggest bugs likely to be commonly encountered in the Australian rainforests are the cicadas. They are most obvious when the males are calling for females while perched on plants. They do not rub their legs as the crickets and grasshoppers do, but have an organ called a tympanum on their side of the body that they flex in; when this pops out it makes a sound. The cicada does this so rapidly that it creates what to our ears sounds like a continuous drone.

Adults are short-lived, surviving for only between a few days and a few weeks, depending on the species. In this time they mate several times and the females lays her eggs into vegetation. Safe underground, the nymphs use their pointed rostrum to pierce roots and suck up the sap from plants. It is unknown how long most Australian species spend doing this, but it is estimated from at least a couple of years to as long as 6-7 years, depending on the species. Most of the 200 species found in Australia are found nowhere else, and the highest diversity of species on the continent is in north-eastern Queensland. On a summer's day in the coastal lowlands, the calling of male cicadas can be deafening. At the peak of their mating when there may be hundreds of individuals flying about, they can be easy to see as they cover tree trunks and fly into people and buildings. However, for most of their time in the rainforest they can be difficult to spot.

Water Striders
Family Gerridae

There are over 30 species in this family in Australia. Water Striders have evolved to live their lives on the surface of water. With their long thin legs they can walk on water using the surface tension, which, on such a small scale, is actually quite strong. They can detect ripples in the water from insects that have fallen in and are trapped in that surface tension, and prey on them. They are most common on still fresh water and can be seen in calmer pools along rainforest streams. Often it is the shadows of the indentations in the water surface that are noticed more than the actual insect itself.

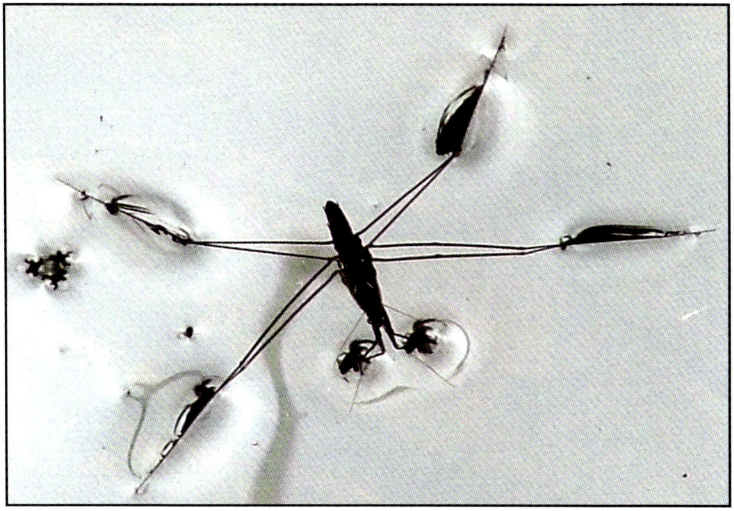

Shield Bugs
Superfamily Pentatomoidea

The Shield bugs are recognized by the large triangle shape (the scutellum) between the wings.

Within this large group are the 'stink bugs' (pictured right, Possum Valley). These little guys often have contrasting or bright warning colouration, and if disturbed they emit a strong metallic smell.

Beetles
Order Coleoptera

Beetles are similar to the Bugs in their rounded appearance and their hard protective forewings that fold over the softer hindwings. However, they differ in the fact that the wings usually meet straight in the middle (to form the protective 'elytra') and they have efficient biting rather than sucking parts. The elytra is often shiny and produces some beautiful colours and textures. This harder layer is lifted up when the hind wings are used for flight.

Whirligig Beetles
Family Gyrinidae

The 'Whirligig' Beetles comprise their own family, with hundreds of species found whirling about in circles on the surface of the quieter waters all around the world. They are in fact looking for dead insects or ones that are trapped in the surface tension of the water. Their eyes are divided into two segments; for looking above the water and below the surface simultaneously. They can be seen in the quieter parts of creeks in the lowland forests, and often end up in the swimming pools of some of the rainforest lodges.

Click Beetles
Family Elateridae

This large family has about 800 species (Zborowski and Storey 1998). They are long, flattened beetles that can be attracted from the forest to lights around lodges. If they fall upside down on a flat, human-made surface, they have the remarkable ability to right themselves by flicking up into the air. This has probably evolved to confuse predators, but it has also turned them into toys for curious kids everywhere.

Scarab beetles
Family Scarabidae

The Beetles of this huge family include thousands of species in Australia, including the pretty Christmas type beetles, the dung beetles and the huge rhinoceros beetles. The Subfamily Dynastinae includes what is perhaps the best known of the tropical Australian beetles, the Rhinoceros, Goliath, or Hercules Beetles. These large shiny insects are mainly noted for the big hook-like protuberances on the head of the male that are used to battle with other males; when the fights take place in trees, the loser may end up falling to the ground. These beetles can give a loud hiss when disturbed. Their hooked, serrated claws give them efficient grip on trees, but also on humans and their clothes. They can often be seen in large congregations when attracted to a shared food source such as when chewing the bark off trees.

Longicorn Beetles
Family Cerambycidae

This large family has over 1200 species (Zborowski and Storey 1998). They can be long, large beetles, with thick, long antennae (pictured below, JCU, Cairns). The larvae are large white grubs that live and chew inside decaying wood, and they are favourite food for various indigenous people around the world. The ability to chew on wood is due to the heavy mandibles, so this is a good reason to be careful of their bite.

Flies
Order Diptera

This order contains many important and ubiquitous species that are also found in the rainforest. Unfortunately, many species in the forest are annoying, such as mosquitoes and March/Horse flies. The bites are from hungry females who need protein rich blood to provide for their eggs.

Mosquitoes
Family Cuclidae

There are over 3000 species of mosquitoes found around the world (McGavin 2000) with some 350 species in Australia (Russell 1999). They have long legs, a thin body and the mouthparts are formed into a long proboscis. Females make a high pitched annoying buzz sound; this attracts the opposite sex, with the males having plumose antennae to pick up the vibrations (Zborowski and Storey 1998). Mosquitoes have aquatic larvae known as 'wrigglers' (Zborowski and Storey 1998).

When a mosquito bites a human, they are inserting a long section called a stylet into the skin to draw out the liquid (Russell 1999). Different mosquitoes attack different species of vertebrates, including frogs, birds and mammals. Some species, such as *Culex* spp., attack mainly birds, while others, such as *Aedes* spp., attack mammals. Some mosquitoes inject a compound through their saliva which reduces blood clotting; this is what often causes the bites to itch later on.

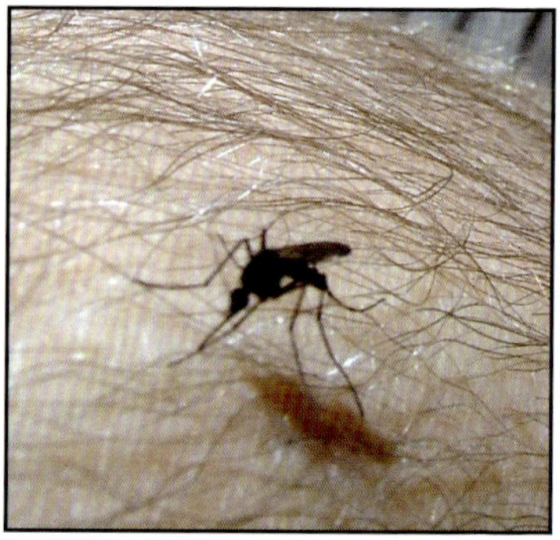

While most species cause nothing but itching, there are some species that spread fatal diseases by introducing parasitic animals and viruses back into the bloodstream. Diseases spread by mosquitoes include Dengue Fever, Yellow Fever and various malarias.

Family Dolichopodidae

Not all flies in the rainforest are bad, for there are also some very attractive species. The over 300 Australian species in this family frequently display beautiful metallic blue, red and green colours (pictured left, Cairns Botanical Gardens). They are also distinguished by their long legs. Most are predators on other insects (Zborowski and Storey 1998).

Moths and Butterflies
Order Lepidoptera

This order contains, and only contains, the moths, skippers and butterflies, and has at least 100,000 species worldwide (Zborowski and Storey 1998), with over 10,500 species described in Australia so far (Braby 2004). The moths make up the majority of families and species in the order. Moths and butterflies range in size from a coin to a dinner plate. There are several characteristics that can be used to define the order Lepidoptera.

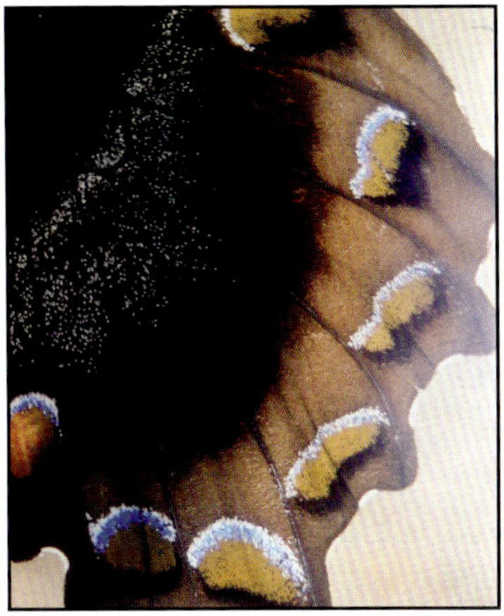

One of the most distinctive features gives the order its name; *lepidoptera*, which is derived from the Greek words for 'scale' and 'wing' (Klots 1983). This is in reference to the many overlapping scales present on the head, body, legs and especially wings of the adult moth or butterfly (Harvey and Yen 1997). The scales can be seen as small dots in the photograph; this is an image of the wing of what is arguably the prettiest of Australian butterflies, the tropical rainforest dwelling Ulysses Butterfly. Contrary to popular belief, these scales are not necessary for flight; they are used for colour in camouflage and communication.

Differences between moths and butterflies

From a taxonomic point of view, butterflies are just several families of moths that have evolved to be more active in the daytime and are thus generally more colourful. There is no single feature that distinguishes these butterflies from all the other moths, but a combination of various features can give clues as to whether the insect should be considered a moth or a butterfly. These include; wing placement when alight (upright in butterflies, flat out in moths), the presence of a frenulum locking in the wings, clubbed antennae (butterflies) or furry antennae (moths), and either diurnal (butterflies) or nocturnal (moths) activity. It is the combination of these features that has to be taken into account, and there are always exceptions. For example, the differences are blurred in groups such as the Hespirid Skippers.

Family Geometridae

Members of the family Geometridae begin life as larvae that we call 'inchworms' or 'looper caterpillars', named because of the distinctive way they travel. They later develop into many of the small and medium sized moths that lay flat against trunks in the Australian tropical rainforest and most other parts of the world. They are often coloured in browns or greens ('emerald moths') to blend in with the tree trunks and the lichen.

Bracca spp.

These are medium to large moths, with striking blue, yellow and white colouration (pictured right, Hypipamee Crater). There are 2 species living in the rainforests of north-eastern Queensland and south-eastern Queensland, the former species is ***Bracca rotundata*** (Zborowski and Edwards 2007).

Dysphania spp.

Four o' Clock Moth *Dysphania fenestrata*. So named because usually seen in afternoon, early evening. One of region's few daytime flying moths; thus has the bright colours of a butterfly. Wings are black and light blue with some yellow on hind wings (pictured below, right, Cairns). Also sometimes called 'Bumble-bee Moth' as has large, fury yellowish body. Caterpillar is quite large; yellow with black spots (pictured below, left, Cairns). Feeds on leaves of *Carrallia*, a small rainforest tree.

Family Uraniidae

This family contains about 700 species throughout the world, with 36 of them recorded so far in Australia (Zborowski and Edwards 2007). The subfamily Uraniinae includes some of the most spectacular, including many large and colourful day-flying moths in the rainforests of Asia, New Guinea and Australia.

Alcides spp.

These day-flying moths are coloured in spectacular oranges and blacks. In size and shape they somewhat resembles Swallowtail butterflies, but they land with their wings spread out like a moth, and usually 'head down'. The **Zodiac moth *Alcides zodiaca*** (pictured right, Cairns Botanical Gardens) is common in the rainforests of north-eastern Queensland, where the larvae feed on the rainforest *carallia.*

Lyssa spp.

These stunning moths are swallowtail in shape and size and are strikingly marked in black, dark brown and white (pictured right, Sepik River). They fly at night around rainforest, from Australia to tropical Asia.

Hawk Moths
Family Sphingidae

The hawk moths are found all over the world, including over 65 species in Australia (Zborowski and Storey 1998), and as with most Lepidoptera there are more species in the tropics. They are often large moths with thick bodies, and are very distinctly shaped like a jet fighter, (which is presumably more aerodynamic than many other moths). The species in this family have a distinctive larval stage, for the caterpillar has a large spine sticking out from the rear of the body, and they are often colourful and have false eye spots. As an adult moth they have a large and obvious feeding proboscis, which they use to poke into relatively deep flowers, hovering as they take the nectar. The moth pictured below is of the genus ***Gnathothlibus***. They are large, handsome brown moths with distinctive white lines along the side of the body, and the larvae feed mainly on grape vines from the family Vitaceae (Zborowski and Edwards (2007).

Family Saturniidae

This family has about 1500 species around the world, however only 15 of these are found in Australia (Zborowski and Edwards 2007). The family includes many spectacular moths, including the Emperor moths. Most species in the family are large, and the body is usually thick and hairy. The large wings usually have a triangular 'window', where the scales are transparent, and the tip of the forewings bend out. The antennae are distinct; they are held out back behind the head, and have many small projections off either side along their length (Zborowski and Edwards 2007).

Hercules Moth *Coscinocera hercules*. Biggest moth of the Australian tropical rainforest, indeed on the whole continent. Further, this species and tropical Asian 'Atlas moth' compete for biggest moths in the world, if measured by wing area. Colours are various shades of leaf browns with (as with many species in this family) small 'windows' in wings. Males have long 'tails' off the wings (pictured above, Ferntree Lodge). The equally large larvae feed off leaves of some rainforest pioneers including Bleeding Heart and Celerywood. Adults are very short lived, generally seen in the wetter months from January to April.

Tiger Moths
Family Arctiidae

Many of the more commonly seen species of 'tiger moth' are coloured orange and black. They are probably most famous for the fact that some species can confuse the insect eating bats that are trying to prey on them by producing their own ultrasound calls as counter signals to confuse or warn the bat of their distastefulness. Species in the genus ***Utetheisa*** (pictured left) also sequester poisonous alkaloids from their host plants to make them distasteful to predators.

Family Aganaidae

This small family includes some colourful species that may be active by day, and are thought to be distasteful to predators (Zborowski and Edwards 2007). ***Agape chloropyga*** is a large bright yellow species that is found from tropical Australia to south-east Asia. The larvae feed on fig trees (Zborowski and Edwards 2007).

Family Noctuidae

With over 35,000 species worldwide, this is the biggest of all moth or butterfly families. It includes many major crop pests, such as the army worms. In the rainforest, this family has species known as 'fruit-sucking moths', as they can pierce fruit to suck the juice out. They are usually brown, but often have hints of orange, white or purple. They may be seen at night on native, cultivated or even discarded fruits (pictured below, Dubuji).

Skippers
Family Hesperiidae

In their features, the Skippers are somewhat intermediate between what is commonly considered the butterflies and the moths. They are usually fairly small, often hairy, and although some have very attractive colours, they are not usually striking. When they land, their wings are held upright, not quite folded up together, but held at an angles.

One of the more obvious and colourful species is the **Banded Demon *Notocrypta waigensis*** (pictured above, Paronella Park), coloured contrasting black and white. The caterpillars feed on various gingers.

Superfamily Papilionoidea

This superfamily includes all the rest of the species that we commonly know as 'butterflies'.

Swallowtails
Family Papilionidae

This family includes some of the bigger and prettier butterflies of the Australian tropical rainforest. The common colour pattern is a background of black with smaller patches of brighter red, green, blue or white.

Birdwings
Ornithoptera spp.

Birdwings are centred mainly in New Guinea, where the group includes species that are the largest of all butterflies. They are also found north into tropical Asia and south into tropical and sub-tropical Australia, where 3 species are found in the rainforests along the east coast. The larvae of the Australian species all feed on leaves of various Aristolochid vines. At some flowering plants, larger individual birdwings dwarf some native birds, such as sunbirds and myzomela honeyeaters.

Cairns Birdwing *Ornithoptera priamus*. Male is a brilliant black and green (lower individual in mating pair in photograph at left), female is a duller black and white with yellow, but larger (higher individual); both sexes have red on body. Best seen in wild around Lake Barrine gardens in summer, or in captivity at Kuranda Butterfly Sanctuary.

Papilio spp.

Some of the species in this genus are known as the 'orchard butterflies' as their caterpillars have been found on cultivated citrus plants. In their natural habitat in the rainforest they live on the other species in the citrus family, the Rutaceae.

Ulysses Butterfly *Papilio ulysses*. Probably the Australian tropical rainforest's most spectacular butterfly, with it's bright blue wings and black markings. When perched, the iridescent blue is hidden by the still attractive but darker underwings (pictured right). Often fly above or on edge of canopy, frequently along rivers and side of road. On sunny day, flashing of bright blue as it flies can be seen from hundreds of metres away. The beauty of this butterfly means it serves as symbol for many businesses along tropical east coast of Australia. Caterpillar feeds on Pink Evodia tree that can be found in tropical rainforest and now commonly planted in gardens.

Orchard butterfly *Papilio aegeus*. Best known and most urban of the Orchard Butterflies. Also one of the larger species on the edge of forest. Mainly black wings, with patches of white on forewing and patches of bright red on hindwing. Male is darker (pictured below, lower photograph), while female is larger in size and has more white on the fore wings and a row of red dots on hindwings (pictured below, upper photograph). As suggested by common name, caterpillar feeds on cultivated citrus as well as other species in that family. In early stages, larvae is brown and white and resembles bird poo; as the caterpillar grows it turns leaf green.

Family Nymphalidae

This family has over 80 species in Australia (Braby 2004). It is a wide and varied group that is better viewed in the different subfamilies.

Bush-browns, Ringlets
Subfamily Satyrinae

The butterflies in this subfamily are usually quite small, and coloured with browns and oranges. At first they appear plain, but closer inspection reveals them to be quite attractive, with many species being decorated with eyespots, especially on the underwings, and this provides their common name 'ringlets'. The caterpillars usually feed on species of grass, thus they tend to be found in grassy habitats. Most species have their distribution in south-east Australia, but several species are found in the tropics, where they can be observed in the grassy edges of the rainforest.

Subfamily Nymphalinae

Eggflys
Hypolimnas spp.

Blue Banded Eggfly *Hypolimnas alimena* (pictured left, JCU Cairns). Striking butterfly, mainly black with blue edges. Seen on the edge of the forest.

Common Eggfly *Hypolimnas bolina*. Sexually dimorphic to extent that many observers assume that female and male are different species. Male (pictured below, left) is beautiful sheeny blackish purple, with large white spots edged in the purple on each wing. Female (pictured below, right) has larger patches of white as well as orange on the forewings. Males territorial, sit on regular perch and fly out to challenge other males, sometimes even other species of butterflies. This behaviour and competition has been the subject of recent research at James Cook University in Cairns (Kemp 2005). Is an open habitat species, but are often seen at edge of rainforest, such as along walking tracks and roads.

Subfamily Heliconiinae

Red Lacewing *Cethosia cydippe*. Large and pretty butterfly with bright red and black on the upper wings when seen in flight. Upon landing, wings fold up and are brown and red underneath, with patches of white, and black edges to wing. Caterpillars are covered in black and yellow bands and found feeding on the native Red Lacewing vine in passionfruit family. Most common in lowland rainforest, and may be seen on tracks through rainforest. Found from Townsville to top of Cape York and also common in lowland rainforests and gardens of New Guinea.

Vindula spp.

Cruiser *Vindula arsinoe*. Large orange butterfly. Male is plain orange (left, upper photograph), while female tends to have more white on wings (left, lower photograph). Usually seen in the lowlands, such as Behana Gorge, around picnic clearing at Mossman Gorge.

Rustics

As suggested by the common name, these are orange-brown butterflies.

Australian Rustic *Cupha prosope*. Medium sized, attractively coloured with border of dark brown around edge of wings (which gives it's other name of 'bordered Rustic'), inside of which is bright, light orange, and inside of that is a darker orange. Bright orange spots on tip of wings. Caterpillars feed on plants in the family Flacourtiaceae. Found in tropical rainforest in Australia and New Guinea.

'Whites' and 'Yellows' Family Pieridae

The species in this family are usually small to medium butterflies. As the common names suggest, many species are variations of simple white or yellow. There are at least 30 species found in Australia, and several of those can be found on the edge of the rainforest or in clearings.

Albatross *Appias* spp.

There are 5 species of this genus found in Australia. They are medium sized and coloured mainly white with various other shades of yellow and black. All the species in Australia, except for the common species below, are restricted to small pockets of rainforests in tropical Australia, centred mainly in Cape York.

Common Albatross *Appias paulina*. Coloured white, yellow and black. Females more striking; white, with yellow on underside of hindwings, and black edges to all wings (pictured below, right). Male lacks much of the strong contrasting black edges (below, left). Caterpillars feed mainly on *Drypetes deplanchei*, and also *Capparis canescens* in north-east Queensland (Braby 2004). Known as 'Common' Albatross as is the most widespread of genus, found from Cape York to Tasmania. Strong fliers and may migrate south in good years (Braby 2004), and can be seen on islands off east coast. Found on edge of rainforest, being common around monsoon forest in Cape York.

Jezebels *Delias* spp.

There are at least 165 species in this genus of butterflies, which is mainly centred in the uplands of New Guinea, where there are species virtually restricted to small mountain top areas (Parsons 1999).

A few species are found in the warmer lowlands, and several of these species are also found in tropical Asia and Australia. In Australia, there are 8 species in the genus. Most species are black and white with splashes of yellow on the undersides of the wing; the Australian species also commonly have blotches of red on the wings. The caterpillar larvae feed mainly on the leaves of mistletoe plants of the family Loranthaceae.

Migrants
Catopsilia spp.

There are 4 species of this genus found in Australia. These small to medium sized butterflies are usually yellowish in colour. They are so named because they travel long distances in large and obvious numbers. The caterpillars feed mainly on legumes such as *Senna* and *Cassia*. They are found in a variety of habitats, including the edge of rainforests.

Lemon migrant *Catopsilia pomona*. Medium-sized pale yellow butterfly with dark edges to wings. Can be seen fluttering around clearings and along road sides in lowland tropical rainforest. The small green, cryptic larvae feed on various *Cassia* including the common tropical garden tree, the Golden Shower Tree *Cassia fistula*. Best known for their mass migrations; most years at the start of summer huge numbers of them appear, flying through Cairns and Cooktown.

Grass Yellows
Eurema spp.

There are 7 species in this genus. Most species are found mainly in grassland and woodland. The caterpillars feed on a variety of plants, with legumes being common, including *Cassia* and *Acacia*.

***Eurema hecabe*.** Most common of genus. Very widespread species, with other sub-species found throughout tropical Asia. Often found in grassy areas on the edge of the rainforest.

Family Lycaenidae

The butterfly family Lycaenidae contains mainly small butterflies often of one colour, and simply named accordingly, such as the self-explanatory 'Blues' and 'Coppers'. During the larval caterpillar stage, many of the species in this family have associations with ants. Some of the 'Oakblues' and 'Azures' are small but beautiful deep-blue butterflies, and can periodically erupt in large numbers in creeks and grassy areas along the edge or in the understorey of the forest.

Wasps, Ants & Bees
Order Hymenoptera

This large order contains some 15,000 species in Australia (Zborowski and Storey 1994). It includes all the ants and the various families of bees, but is really dominated by the wasps. Despite the differences, wasps, bees and ants share a typical body shape that is elongate, with long wings, and has a pinched in 'waist' between the abdomen and the thorax.

Although many species are solitary, quite a few groups have developed complex colonial systems. The ants are probably best known for this, but it has also evolved in many species of wasps and bees. When it does occur, there is usually one main queen, who is protected by the rest of the colony. Both those protectors (the 'soldiers') and the makers of the nest (the 'workers') are female. In many species the female has modified her egg layer, the ovipositor, into a special organ of attack and defence; the dreaded sting!

Wasps

There are many different families of wasps represented by species in the Australian tropical rainforest, from the tiny fig wasps to the bright metallic green cuckoo wasps. Adults of many species are known for their highly predatory nature (pictured below, Papua New Guinea).

Family Vespidae

This is likely the most infamous of the wasp families, for it includes mud wasps, most of the colony constructing paper wasps, as well as the introduced and aggressive European wasp. With over 324 species in Australia (Zborowski and Storey 1998), the family is divided into various sub-families.

Paper Wasps
Sub-family Polistinae

This subfamily contains the Paper Wasps, well known for their sometimes large and elaborate nests. They are amongst the most obvious of wasps seen in the tropical rainforest. Individuals of the colony can be seen crawling on the outside of their paper or mud nests, often built onto tree trunks, hanging in branches and leaves, or on walls of lodges in the rainforest (pictured left). While the adults drink nectar, they kill and catch caterpillars for the predacious larvae. This prey is subdued with their stings and they also use this weapon to protect the colony, so it is always a good idea to avoid the nests.

Bees

Most of the native bees in the tropical rainforest are small and stingless.

Family Trigonidae

This family is found in the rainforests and woodlands of the tropical Americas, Asia and Australia. The trigonid bees are small, dark bees that resemble flies. They can sometimes have the same annoying habit of hovering around humans to drink the sweat off their bodies, and thus one of their common names is 'sweat bees'. They usually make wax nests in tree hollows, and in the rainforest these nests can sometimes be seen on outside wooden tables and buildings.

Ants
Family Formicidae

Ants have to be one of the most diverse, interesting and ecologically important families of animals in the world. They are of course found all over the globe, with the greatest diversity in the tropics. The family is split into several different sub-families. Australia has over 4000 species (Zborowski and Storey 1994).

Ants have highly evolved social structures, with individuals developing into different 'castes' that have different functions, such as workers or soldiers. Generally, the regularly visible ants such as these soldiers and workers are sterile females. The winged flying males only appear during the times of reproduction and establishing new nests. Studies in the last few decades have revealed the efficiency of their communication through the use of pheromones.

Most adult ants actually feed on liquid foods such as nectar and fruit juice. The mandibles seen in adult ants are for defence and tearing apart food. The food that we often see adult ants carrying is usually being taken back to the larval ants, who are in the form of tiny grubs, and much more predaceous than the adults.

Polyrachis spp.

There are almost 500 species in this genus, and they are found throughout the Old World tropics. Although there are a hundred described species in Australia, there is probably another hundred yet to be named (Anderson 2000). The genus name *Polyrachis* refers to the many hairs on the ants abdomen. In some species these hairs create a often a golden tinge, and thus they are commonly referred to as 'golden bum ants'. They are found in every terrestrial habitats, including arid shrublands, savanna woodlands, mangroves and rainforest.

Spider Ants. These remarkable and distinctive ants are found walking alone along the floor of the tropical rainforest. As suggested by the common name, they do at first glance superficially resemble spiders. But a closer look reveals the typical 'pinched-in' waist of the order. They are large, with long white tipped legs and very long antennae.

Weaver Ants, Tree Ants
Oecophylla spp.

Weaver Ants are found in the tropics of Africa, Asia and Australia. There are now 2 recognized species with several different races, including the 1 species found in Australia as discussed below.

They are large ants, usually with a brownish body and an abdomen that varies from brown to reddish to bright green. They have large and effective mandibles. Weaver ants feed their larvae on a large variety of foods, ranging from fruit, to dead animals, to catching their own insect prey. They are quite ambitious predators, for example catching onto relatively large, struggling butterflies with their strong mandibles. The generic name of *Oecophylla* means 'house-leaf' and refers to the fact that these ants make a nest out of leaves (while still attached to the tree); this is something very few other ants do. The building of the nest is an interesting process. The establishment of a new colony is begun by the queen ant (upper picture). She lands on a suitable leaf and discards her wings (Serventy and Raymond 19--). She then produces several larvae and uses their silk to build a structure to house the developing workers. The first adult workers then proceed to build a nest. Fresher leaves are usually selected because they are flexible and can be bent. The ants may form bridges or chains with their bodies to bring the leaves together. Other ants then grab their larvae and squeeze silk out of their bodies like toothpaste from a tube. As the silk dries, it pulls the leaves together into a large, football-sized nest. The ants vigorously defend this nest, grabbing flesh with the mandible, then bending the abdomen over and squirting formic acid into the wound.

This results in a short, but sharp, sting that is frequently encountered in tropical Australia. And this is how many people first interact with the local species, the **Green Tree Ant *Oecophylla smaragdina***, probably the most commonly encountered ant in the Australian tropics. However, you can also bite back; as these animals are edible, and in fact quite tasty.

Fish

In an international comparison, Australia has a very low diversity of fish fauna. This is likely due to it being the driest of inhabited continents, with a lack of year round fresh waterways, and the least amount of run-off of any landmass (Allen *et al* 2002). The wettest part of the country, North-east Queensland, is an unusual contrast, with a high and constant rainfall, and a range of rivers and streams. This is reflected in the region having the highest diversity of fish on the continent, with some 130 species; about half of Australia's total (Pusey *et al* 2004). Of course, many of these are inhabiting the wetlands and mangroves, but many fish are also found in the waterways within the tropical rainforest.

Class Osteichthyes

Order Anguilliformes

Eels are true fish, although they are generally considered primitive. They have elongated bodies and tend to have sharp teeth. Over 650 species can be found in the oceans and freshwaters of the world.

Freshwater Eels
Family Anguillidae

There are some 15 species of eel that live most of their adult life in freshwater streams and lakes. However, they all begin their life in the ocean, and then eventually make their way up into freshwater, sometimes crawling across land in the process. There are 3 species commonly recorded in the freshwater rivers and lakes within the Australian rainforest, and they can be seen at Paronella park and Lake Barrine.

Catfish
Order Siluriformes

This group includes the catfish. They are easily identidied by their cat-like whiskers, the barbels. All of the species in Australian waters have venomous spines from the base of the fins and therefore handling should be avoided. This group is most diverse in the freshwaters of North America, but there are several species in 2 families found in the freshwater streams of the Australian rainforest.

Fork-tailed Catfish
Family Ariidae

There are about 80 species in this family around the world, but they are centred in the tropical Australian/ New Guinea region (Allen *et al* 2002). These catfish have, as suggested by the common name, standard fork shaped tail (pictured below) like most other fish but in contrast to the eel like tail of the other catfish family here. They have 3 pairs of barbels at the mouth (Allen *et al* 2002).

Eel-tailed Catfish
Family Plotosidae

There are over 30 species in this family spread throughout the estuaries and coastal waters of the tropical Indo-west Pacific (Allen *et al* 2002). These catfish have, as suggested by the common name, eel like tails. They have 4 pairs of barbels at the mouth (Allen *et al* 2002). Species in this family are also commonly referred to as 'Tandans'. Several species are noted to have elaborate parental behaviour, excavating nests out of the sand at the bottom and looking after the eggs.

Order Atheriniformes

Rainbow fish
Family Melanotaeniidae

There are over 60 species in this family, and more are still being discovered (Allen *et al* 1994). They are only found in the freshwaters of Australia and New Guinea and Australia, and on the latter continent tend to be tropical in distribution.

Rainbowfish are not quite as colourful as suggested by the common name, but they have a subtle beauty, long admired by aquarists (pictured above). They tend to be the most abundant and obvious small fish in most tropical streams. Males are more colourful and deeper bodied (pictured previous page), than females (Allen *et al* 2002).

Order Scorpaeniformes

Family Scorpaenidae

This infamous family includes many strangely elaborate and venomous fish. They tend to hover around or sit on the bottom, feeding on small fish or crustaceans. They generally rest during the day and feed at night. As well as being bottom dwellers, some are very well camouflaged, and thus they are not often seen. Hundreds of species inhabit (and can be trodden on) in various freshwater, estuarine and oceanic environments around the world, including the attractive but dangerous 'Butterfly Cod' or 'Scorpionfish' of the coral reefs.

There are several freshwater and brackish species known as 'stonefish', and these are the most venomous fishes in the world. This group includes the notorious **Bullrout *Notesthes robusta*** (pictured right), which lives in freshwater streams along much of Australia's east coast.

Order Perciformes

Flagtails
Family Kuhliidae

Jungle Perch *Kuhlia rupestris* (pictured left). Probably most obvious of larger fish seen in waterways of Australian lowland tropical rainforest. Can be a very pretty silver, gold or olive brown, with dark-edged scales that appear as spots all over the body. Two dark spots on their tails. If not hunted, can grow to 30 or 40 centimetres. Can easily be seen in faster fresh waterways, such as in pools along the Mossman River, waiting for handouts.

Archerfish
Family Toxotidae

Archer Fish are found in the fresh and brackish waters of tropical Asia, Australia and the Indo West-Pacific. They have a distinctive shape, being laterally compressed, but with a flattened top edge.

They usually feed on invertebrates that fall into the water, being well known for their ability to catch their prey even outside the water's edge. Some species leap out of the water to grab their prey, while others squirt out a stream of water to knock the prey down. They can be observed in Lake Eacham. The most commonly seen species in Australian tropical freshwater is the **Seven-spotted Archerfish *Toxotes chatareus*** (pictured right).

Family Scatophagidae

This tiny family contains only 2 species, centred in the tropical Indo-Pacific (Allen *et al* 2002). They are referred to as Scat or Butterfish, and are silver with stripes and spots, and laterally compressed like a butterflyfish. The juveniles are brightly coloured, and look even more like miniature butterflyfish. They are essentially estuarine fish, but often come up into lower reaches of freshwater. They are considered both attractive aquarium fish and good eating (Herbert and Peeters 1995).

Sub-order Goboidei
Gudgeons
Family Eleotridae

There are over 150 species of gudgeons and they are found all around the world, especially the freshwater and estuarine waters of the tropical Indo-Pacific (Allen *et al* 2002). These generally slow moving fish usually hang near the bottom. They have a long cigar shaped body, a torpedo like head and several species have reddish spots.

Frogs

Class Amphibia

The only types of amphibians found in Australia are the frogs, as we lack salamanders and newts.

Frogs
Order Anura

Frogs are the tailless amphibians. They have two shorter front limbs and two very long hind limbs, a body plan that facilitates jumping very effectively. This design has obviously been successful as it has changed little over many millions of years.

Most frogs have excellent vision, being able to judge distances and detect movement; the former is important for animals that leap about, and the latter is essential for animals that have to catch fast moving prey with their tongue! Their 'ears' are represented by round shapes called tympanums on the side of their head. These are often obvious, and the markings around this organ often assist with species identification.

Most frogs have fairly permeable skin and are capable of some breathing through their skin; but to be able to respire like this requires that they stay relatively moist. This permeability also means that amphibians can 'drink' through their skin, and apart from only 1 species, frogs are not known to actually drink water. The permeable skin can even extract moisture from seemingly dry substrates such as sand.

Within their skin, many frogs have different glands that serve several purposes, such as exuding poisons. This is well developed in many toads, and in the 'poison arrow' frogs of the Neotropics. Frog skin also has the ability to change colour, sometimes quite quickly, due to the presence of 'chromatophores'.

Life cycle

In most frogs, a signal produced by the male has evolved that serves to attract females. This is the well known vocal call, and it is allows for species-specific identification and location, both for the female frog and the herpetologist. These calls can be recorded and displayed as an oscillogram, which displays signal amplitude in volts/millivolts, or a sonogram, which display frequency (pitch) (Barker *et al* 1995). The female frog often has hearing which is tuned into the frequencies of which the male of its own species tends to call; this eliminates confusion with all the other frog calls (Alford 2002).

Many frogs have evolved to develop large vocal sacs. These are blown up to extraordinary proportions to make calls (pictured). Many frogs lack this feature, but may still call. Others may take advantage of the acoustics within piles of rocks or pipes to amplify their call. Often, many individuals will call from the same area, their combined calling perhaps attracting females from some distance.

While rain is often associated with these calls, the exact cues of many species to call and then mate are quite difficult to determine, and may be a combination of rainfall, temperature, humidity and social cues.

Sexual reproduction in frogs is remarkably diverse. It usually involves the male grabbing the female in a sexual position called amplexus (pictured left). Many male frogs may develop a hard patch of skin on the back of their thumb; this is presumed to aid in his grip on the female during this position. The female frog is generally larger than the male, sometimes substantially so. This enables her to support him during mating, especially if she is swimming or climbing. As she releases the eggs, he fertilizes them externally.

Eggs of frogs are usually laid in a moist or watery environment. These soft, jelly-like eggs then usually hatch into aquatic larvae called tadpoles. Over various stages the larvae will then undergo transformation into a more terrestrially adapted adult.

There are variations on the basic frog reproductive cycle in species found in the Australian tropical rainforest. The Stony Creek Frog *Litoria leseuri* actually builds a nest separate but close to the creek to mate and lay her eggs. Some frogs have developed reproductive cycles that rely less directly on water; the Barred Frogs *Mixophys* actually lay their eggs out of the water. They are, however, near the edge, so the tadpoles are eventually washed in (Barker *et al* 1995).

Some frogs have developed reproductive cycles that can occur even further from the water. The microhylids have the unusual reproductive strategy of laying eggs that hatch directly into little frogs; they skip the usual larval stage of the tadpole. The eggs do not dry out in the moist leaf litter of the rainforest. Some species of microhylid, such as *Spheophryne fryi,* have been observed guarding their nests of eggs (Alford 2002).

Declining frog populations

In the last few decades in different parts of the world, including the rainforests of Australia, certain frog species have experienced drastic declines in what had been thought of as relatively pristine rainforest streams. Many different causes have been implicated (see research on frogs under the Humans section later in this book). A fungus has now been implicated, and identified as *Batrachochytrium dendrobatidis*. It has dermal and systemic effects on the infected frogs.

Frogs of the tropical rainforest of Australia

The Australian tropical rainforest, being hot and wet just as frogs like it, has a good diversity of interesting species. Frogs are often used as a symbol in tropical rainforest around the world, and the 'Wet Tropics' of North Queensland have certainly followed this lead. Because of this advertising, visitors often wonder where all the frogs are; however, they require patience and getting a little bit dirty to find them. The actual interior of the rainforest can sometimes seem quite poor for frogs, with more species on the edge and associated ecosystems, such as adjoining open woodlands and wetlands. Fortunately, some may come to you, for certain species have urbanized quite well and many frogs can be found living happily on boats (pictured below) and inside lodges, drain pipes, and even toilets in, around and on the edge of the rainforest.

Tree Frogs
Family Hylidae

The Hylids are very attractive and photogenic frogs and they are quite often featured in books, magazines and on T-shirts. They are found in many different environments all over the world.

There are more than 790 species in 40 genera (Cogger and Zweifel 1998). Australia has 3 genera and over 70 species. The highest diversity of species of tree frogs is found in the wetter forests of Australia, New Guinea and most especially tropical South and Central America, where about two thirds of the species of this family occur. This latter area also is most likely where the tree frogs originated, with a fossil record going back to more than 50 million years (Cogger and Zweifel 1998). Hylids are generally absent from the polar regions and much of Africa (EOTAW 1977, Cogger and Zweifel 1996); however, in Africa there is another closely related family that replaces them.

The family name of 'Tree Frogs' is a general one and a bit misleading. There are many species of frogs in various other families that can climb well and can also be found in trees. Equally, there are also members in this 'tree frog' family who may live in seasonally dry habitats and do not climb trees, such as the large and ground dwelling *Cyclorana*. Generally, however, the true tree frogs are adept climbers, jumpers and clingers, and have a suite of features that unite them as one family.

Physical characteristics

Tree frogs generally have long limbs, rounded toe discs and large mouths. The eyes tend to be laterally placed on the head and directed forward (Cogger and Zweifel 1998), presumably to be effective in judging distances when leaping and catching their prey.

Like other tree climbing frogs in other families, there is often a flattened area of round skin called a disc on the end of the digits. The discs secrete a sticky substance from mucus glands in the skin (EOTAW 1977). If the toe pads are examined under a microscope one can see interlocking irregularly shaped cells with narrow gaps; these gaps catch onto tiny irregulations on the climbing surface, even seemingly smooth surfaces such as glass.

They also have an extra disc-shaped piece of cartilage in each digit of the fingers and toes which allows the disc to be pressed up against a surface even when the finger or toe is at an angle (Cogger and Zweifel 1998). The detailed pattern of irregularly shaped cells can also occur on the skin of the underside of the stomach (Martin 199-), which explains why many frogs squeeze their stomach against a window; they are increasing the surface area to enable themselves to cling to the smooth glass.

Most tree frogs need to return to water to breed, with nearly all species laying eggs in water. Subsequently they have a free swimming aquatic tadpole stage. The tadpoles of many hylid species tend to have laterally placed eyes, to see below and above the waterline (Cogger Beonos and Zweifel 1998) and feed by filtering microscopic food particles (Cogger and Zweifel 1998). There are, as usual, exceptions and variations to the standard tree frog life cycle. Some species may lay their eggs on vegetation overhanging water, so when tadpoles hatch they plop into the water below (Cogger and Zweifel 1998). Others may breed in holes or vegetation that holds water, such as epiphytic ferns or orchids (Cogger and Zweifel 1998).

Lacelids
Nyctimystes spp.

Most of the 20 or so species of Lacelids are found only in New Guinea; another one to three (depending on classification), are found in the Cape York region of Australia. Within this range, they are normally found in vegetation and on rocks around fast flowing streams in rainforest. The most distinctive feature of these frogs are their big beautiful black eyes that are so photogenic. The eyes of this group of frogs have another diagnostic feature in the striking lace-like venation on the eyelid that is visible when the eyes are shut; this pattern is called 'palpebral venation'. Its appearance thus results in the common name of 'Lacelid' for this genus. Barker *et al* (1993) suggests this probably helps in camouflage when the frog has its eyes closed during sleep or rest. Unlike many other frogs, when the pupil contracts it forms a vertical rather than horizontal slit. As most Lacelids live near rainforest streams, they breed in fast flowing water; the eggs are large and have the tendency to clump together and thus not get washed away (Nightingale 1992). Tadpoles have a strong sucker mouth to cling to rocks and vegetation to avoid being swept away downstream (Nightingale 1992). They are also thus strong swimmers, with a flattened body, low tail fins, dorsally placed eyes (Cogger and Zweifel 1998) and a muscular tail (Freeman 1998).

Australian Lacelid *Nyctimystes dayi*. Big, black eyes, vertical pupils and 'lacework' eyelids. Variably coloured upper surface, but usually golden brown, often with lichen like spots. Whitish underneath, with very fine lappets on edges of back legs. Usually found on vegetation or rocks around fast flowing streams in tropical rainforest of north-east Queensland. Currently appears more common at lower altitudes, but this may be due to the recent declines in stream dwelling rainforest frogs. Was only discovered and named in Australia in 1969 (Tyler 1994). Currently considered as the only one of its genus found in Australia, but prior to 1987 it was considered to be one of two or three species; and this may change back again.

Litoria spp.

This is the largest genus of frogs in Australia. They are generally slim, long-legged frogs that live on vegetation, but there are also species that live in streams, wetlands, and on the ground.

Orange-thighed Treefrog *Litoria xanthomera*. Very pretty slim frog, green with yellow-orange underparts and orange-red eyes. Unlike next species, there is some green on lower lip. Thigh colour is a deep orange. Usually only seen and heard on very wet nights when comes down from canopy to call; a drawn out 'waaa...waaaa....' similar to next species, but with quiet trill on end. Lives in rainforest canopy North-east Queensland.

Dainty or Graceful Treefrog *Litoria gracilenta*. Similar to previous species. Very pretty slim frog, green with lighter yellow underparts and lighter orange eyes than *L. xanthomera*. There is often a yellow line over and behind typanum. Hidden thighs a distinctive purple'ish. The skin is much more granulated than in the previous species. By day often seen out and exposed, but in a distinctive position with the limbs tucked away and stuck on vegetation or the glass of a window. On wet nights a drawn out 'waaa...waaaa....' call can be heard. Only on edge of rainforest; more common in woodlands and swamps, and also urban situations, such as sides of roads, fields and in gardens. Mainly lowlands, from northern NSW to Cape York.

Dwarf Tree Frog or Green Reed Frog *Litoria bicolour* and *Litoria fallax* Cute, tiny green frogs. Two similar species. *L. bicolour* can be distinguished from *L. fallax*, by barely discernable pale stripe along side where green upperparts and cream underparts meet. Call is a clicky "criiiiick.....pip". Usually found perched nimbly on green reeds in water, including ponds or lakes on edge of rainforest, eg; Lake Barrine.

Green Tree Frog *Litoria caerulea*. Large frog, usually green, but can range from dull olive brown to bright green, and even occasional blue specimen. May have small pale spots on back. Underparts white and granular. Fully webbed feet and toe pads. Often a large fold hanging over tympanum in larger individuals. Call a deep "crrawwk", often made over and over; can be loud if amplified by drain or pipe. If caught by predator, gives horrible painful cry. Found throughout much of Australia, including the centre, the east, and most common in tropical top third of Australia. Happy to live with humans, found in pipes, letter-boxes and is frog most likely to be seen in toilet. Humans also enjoy frog; is a popular pet, especially overseas, where may be called "White's Frog" or "Lumpy Green Tree Frog".

White-lipped Tree Frog *Litoria infrafrenata*. One of most obvious and common of rainforest frogs. Females can grow to enormous sizes, and is probably biggest tree-dwelling frog in the world. Similar looking to Common Tree Frog, but larger, less 'lumpy' and distinguished by 'white-lip'. Call is a strange double 'kurchunk' call, which when given by a group, sounds like a machine factory. Like common tree frog, also gives out a horrible scream when caught by a predator. Common on edge of forest, and even in urban areas such as around Cairns. Can be seen around many of the lowland lodges, and on Daintree River cruises, etc.

Green-eyed tree frog *Litoria genimaculata*. Cryptic frog, exquisite and variable. Body coloured in lichen-like browns and greens. Common name refers not so much to green in the eyes, but to a strip of green that is found on skin above eyes; 'green eye-browed tree frog' would be more accurate name. Has small lappets of skin on back of legs which probably break up the outline and allow to blend in on the rocks upon which it usually perches. Call a subtle tapping, usually drowned out by their small rainforest creek habitat. Occasionally makes its' way into lodges and homes within rainforest.

Common Mist Frog *Litoria rheocola*. Small frog. Usually coloured brownish or tan above and whitish below, often slightly spotty. Often lighter patch from snout which ends abruptly between eyes to a darker brown. Angled snout, large dull orange eyes, webbed toes, large toe pads. On warm, wet nights their soft but persistent 'wreeeek' calls can be heard from creeks. Usually found on boulders (and occasionally surrounding vegetation) in and around freshwater rainforest creeks, seems especially common around lowland rainforest creeks. This is one of the 'disappearing frogs' of the 1980's.

Waterfall frog *Litoria nannotis*. Robust frog with thicker forelimbs. Very restricted range and habitat; only found only in splash zones around waterfalls or large rocks, on freshwater creeks through tropical rainforest in North Queensland.

Stony Creek Frog *Litoria lesueri.* Medium size. Seems to be 2 forms over it's long range along entire east part of Australia. Tropical Queensland form smaller. Coloured from leaf brown to bright yellow in breeding males. White undersides with black and either blue or yellow 'flash colours' on back of the legs. Darker brown stripe from snout, through eye and over distinct tympanum (ear). Call is a very quiet trill, so not often found by sound, but stumbled upon accidentally. North Queensland form has interesting reproductive behaviours: after mating underwater, makes small circular nest in sand and leaf litter at side of stream, lays eggs inside (Barker *et al* 1995). As suggested by common name, often found along stony creeks beds, gullies, rocky dirt roads and tracks in tropical rainforest.

Roth's Tree Frog or Northen Maniacal Cackling Frog *Litoria rothii*. Medium size. Usually grey or brown, but with ability to change colour depending on the situation. Consistent colour found on back of thighs; bright yellow with black patches. Call is almost identical to it's relative and temperate Australian zone equivalent, the Peron's Tree Frog. Both species have a call described as a 'maniacal cackling', or like a sheep's bleating. Can be found anywhere with water in the tropics and often found in sugar cane and other fields, and backyards on edge of rainforest. Can be found in houses and outside toilet blocks at camping grounds, etc.

Rocket Frog *Litoria nasuta*. Medium-sized, extremely streamlined frog, with proportionally very long legs, and angular head. Usually coloured light brown, with various darker brown upraised stripes and spots on back. Call is series of scratchy, staccato notes that begin tentatively, as if frog unsure to call, then continues quite loudly. Found throughout top end and along east coast down to Sydney, sometimes on edge of forest. They are not called 'rocket' just because of shape; they can jump far and fast, apparently making the greatest leaps of any Australian frog. Just try catching one!

Red Tree Frog, Naked Tree Frog *Litoria rubella*. Small frog. Longish body, shorter limbs, distinctly blunt snout. Extreme variation in colour, from reddish brown to translucent so organs can be seen through the skin and thus appears 'naked'. Dark stripe from snout, through typanum and along side of body. Call a consistent bleating, can be deafeningly loud, especially when amplified in drain or pipe. One of Australia's most adaptable and environmentally diverse frogs, as found in range of habitats, from desert to woodland to monsoon forest and edge of rainforest, throughout tropical, eastern and central Australia, and southern New Guinea. Also found in houses and toilets.

Southern Frogs
Family Myobatrachidae

This family dominates the terrestrial and freshwater frog fauna in Australia and New Guinea. There are over 120 species in 23 genera (Behler and Behler 2005). They show the highest diversity of forms and niches of the Australian frog groups, with species adapted to living in arid, high montane, woodland and forest environments. They are best known for the species living in wetlands, such as the Marsh Frogs, that are common in the human populated south east. They are missing only from the arboreal (tree) habitat, a niche filled by the tree frogs of the family Hylidae. Most species tend to be generally more common in the open woodlands and wetlands, and thus they are poorly represented in the Australian tropical rainforest.

Marsh Frogs
Limnodynastes spp.

Striped Marsh Frog *Limnodynastes peronii*. Medium sized frog. Coloured light brown with darker brown blotches and lines going down it's back, including a distinct central cream stripe down the middle. Call is a single note that sounds like a tennis ball hit by a racquet: a soft "bwuck" (if between two males calling, you may be forgiven for thinking you're at a tennis match!). Call as they float in the water, usually near edge hidden by vegetation. If disturbed, may submerge. Long distribution along east coast of Australia, from south-east all way up to Cape York.

Great Barred Frogs
Mixophyes spp.

This is a large brown frog with striking barring on the legs and large dark eyes. The beautifully subdued but varied dead leaf browns of their skin enable these frogs to blend in with the leaf strewn floor of their habitat. This habitat tends to be the rainforest and wetter sclerophyll forests of the east coast. One species lives in New Guinea, 3 in northern NSW and southern Queensland, and 1 species in North Queensland.

Northern Barred Frog *Mixophyes schevilli*. Call is a deep "waaarhhhhk". Has habit of lowering its head and lying flat against ground when disturbed, likely trying to blend in with the forest floor upon which it usually sits and resembles. Has largest tadpole of any frog species in Australia. Lives in rainforests of North Queensland, appears more common at mid and higher altitudes. Because of large size, toad-like shape, and habit of being found on side of roads through rainforest, are often mistaken for cane toad, and deliberately killed by people.

Nursery Frogs
Family Microhylidae

The most cryptic group of frogs in the Australian tropical rainforest are the 'Nursery frogs' or 'Narrow mouthed frogs'. Most members of this family are tiny and dull coloured, and thus very rarely seen. However, their calls are commonly heard on warm, wet nights in the Australian tropical rainforest, although many people may not immediately recognize these often strange calls as coming from frogs, as many species have almost bird-like whistling calls. Of all the species in this family, only the one below is likely to be seen.

Ornate Frog *Cophixalus ornatus*. Very tiny brown frog, size of 10 cent coin. Small lumps on back. This species of microhylid has straight edge to toe pads, and thus triangular-shaped toe pads. Has high and quite loud 'meeeep', like a little toy car, usually calling from perched on tree trunk, often only a few metres from the ground, but difficult to locate.

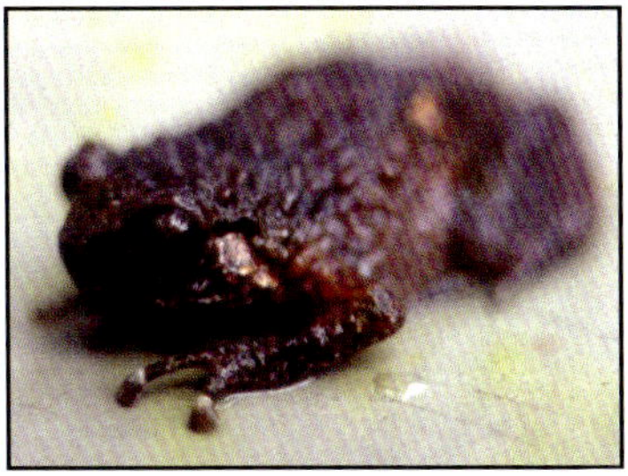

Toads
Family Bufoidae

Toads are a well-known, almost world-wide family of frogs. However, they did not have any representatives in Australia until recently.

Cane Toad *Bufo marinus*. Large. Ugly. Has two large glands, and numerous small ones, all over its skin that release a toxin that deters predation. Because of this, they are one of the few frogs that are likely to be seen during the day, and are one of the most common ground amphibians around the Australian tropics in the wet season. Pleasant rolling calls sounds rather like a generator or a motorbike in the distance, often heard on warm, wet nights. Was introduced into Australia from South America, via Hawaii.

Bull Frogs
Family Ranidae

The classic 'bull-frogs' of Europe and North America are often referred to as the 'true frogs'. There is only 1 species in Australia. The **Wood Frog *Rana daemeli*** gives away its presence by it's gentle croaky chuckle along the slower, smaller freshwater creeks in the lowlands across northern Australia.

Turtles, Lizards & Snakes, and Crocodiles

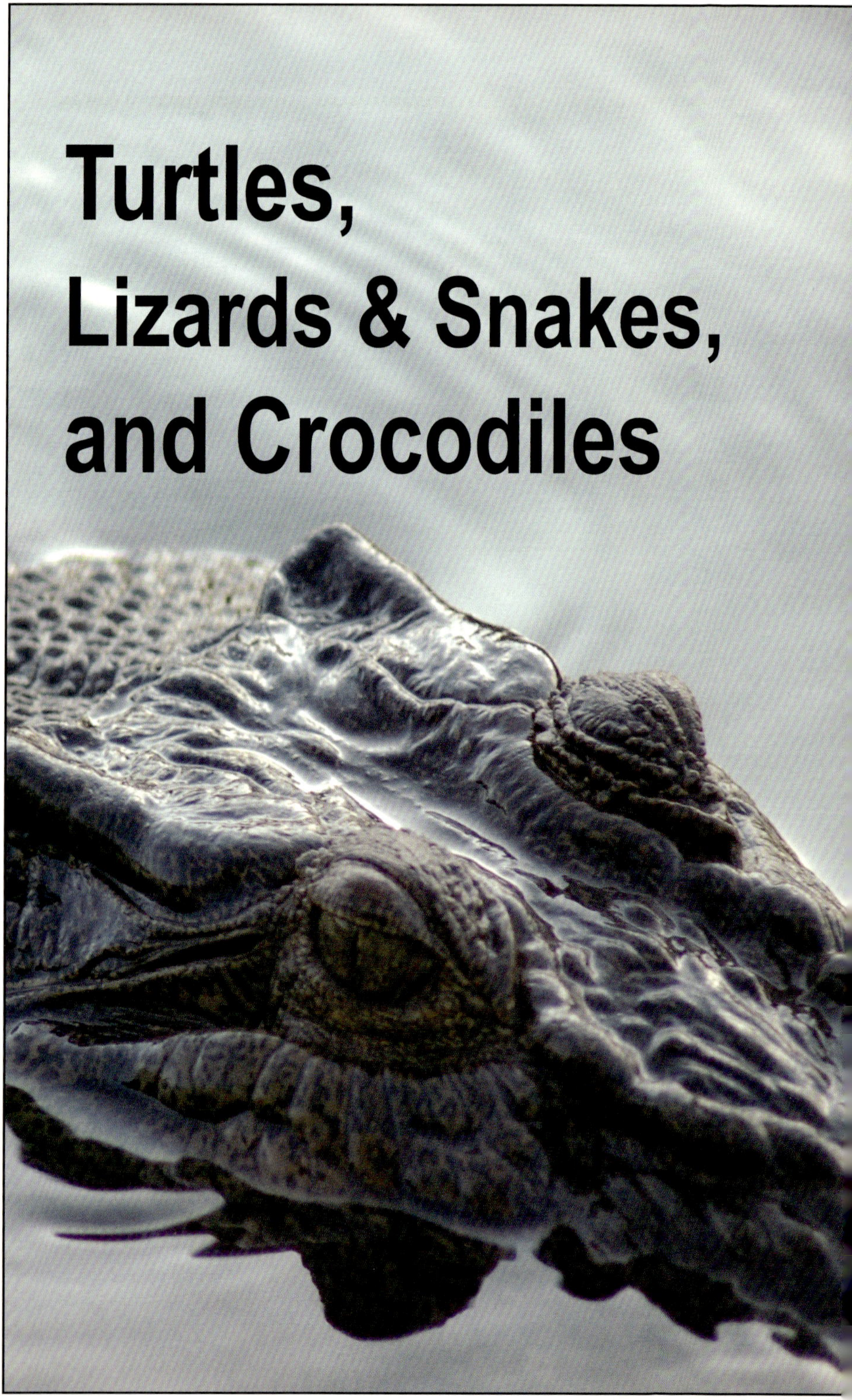

Taxonomy of the "reptiles"

This section groups together what have often been considered the 'reptiles'. For evolutionary and taxonomic reasons, this has been known to be a false group for a long time. The way to represent a true group would be to either enlarge it into a huge and varied class that also included the dinosaurs and the birds, or split the 'reptiles' into smaller distinct groups. The former is not suitable given the importance, size and distinctiveness of the Birds (and dinosaurs) as a class in itself. Therefore, the following section will take the latter path and represent the following as distinct and true groups: the turtles, the lizards & snakes, and the crocodilians.

Evolutionary success

The evolution of reptiles was an important step for animals as they invaded dry land. Amongst other important adaptations that enabled them to do this was their impermeable skin, in contrast to the amphibians, that prevents desiccation of the body fluids, as well as affording some protection against other animals. Some reptiles, such as crocodiles, go even further and have sections where the bone directly supports the skin, forming 'osteoderms'. Turtles have developed a hard layer of 'scutes' atop the extension of their ribs that comprises their shell. This is still part of the skin, however, and a turtle can thus bleed from damage to the shell.

Other important factors in enabling the reptiles to invade land were elements of their reproductive cycle. Males have a penile organ (or two), to transmit sperm directly. After successful mating, the female usually digs a hole into a substrate or builds a nest into which she will lay her eggs. These eggs were different than in the animals that had come before, and were another important step in colonising land; they are covered with a harder shell that is secreted by the mother as it develops through the body. Within this exterior casing is contained a separate environment in which the embryo can develop. However, in some reptiles, females give birth to live young.

Turtles
Order Testudinata

The names 'turtle' and 'tortoise' are only common names and have no real biological significance. They mean different things to different people in different cultures around the world; basically any 'reptile' with a shell is a turtle. There are about 250 species, and they are found from rainforests, to savanna woodlands, to deserts, to rivers and oceans. Turtles are distributed throughout the tropics and also much of the subtropical and temperate parts of the world. They are considered the oldest of the living 'reptiles'. The earliest fossils have been found from the Triassic period about 230 million years ago, at the time of the earliest dinosaurs, and they have changed little from this ancient but successful design. Today, there are generally two types of turtles, based on how they pull their head into their shell, which is in turn dictated by their morphology. The 'hidden-necked' turtles of the suborder Cryptodira can pull their head all the way into their shell; they do this with a vertical bending of the vertebral column (Obst 1998). The 'Side-necked turtles' of the Suborder Pluerodira retract their neck with a sideways motion, their head not necessarily enclosed in the shell. The Cryptodirans include all the marine turtles and many of the land and overseas freshwater turtles. In Australia they are represented by the marine turtles and the 'pig-nosed turtle' of the northern tropical wetlands. All of the other freshwater turtles of Australia are Pluerodirans, and other members of this group are found in central and southern Africa and South America (Obst 1998). Australia completely lacks any true land turtles (often called 'tortoises' overseas).

The shell of the turtle is it's most distinct and important feature. The top half is called the carapace, the bottom half is the plastron, and where the two join there are gaps for the tail, legs and head. The feet and toes vary with the type of turtle. Marine turtles have evolved 'paddles', the heavier land tortoises large clumped feet, while the aquatic turtles, such as those in the waterways of the Australian tropical rainforest, are somewhere in between; they have separate digits with webbing (pictured below). The shell itself is made up several layers, which give it strength. The underlying layer is made up of large, spreading extensions of the ribs. There are several layers above that, with the top layer made up of scales of keratin referred to as 'scutes'. The scales determine the distinctive patterns and colours of the shell, and the differences can aid in recognition of species.

Turtles usually dig a hole in the substrate to lay their eggs; even the marine and freshwater species have to do this, which means they need to climb up out of the water to do so. After laying, there is no more parental involvement, and the hatchlings must dig their way out and immediately fend for themselves.

Family Cheluidae

This family is found in South America, and New Guinea and Australia. They are usually small to medium-sized turtles, with long necks, which enables them to grab a breath from the surface without revealing themselves. They are excellent swimmers and some species can stay submerged for long periods due to the increased vascular linings in their cloaca which can extract oxygen from the water (Wilson and Swan 2003); in other words they can breathe through their bum! Diet varies between and within different species, ranging from aquatic vegetation to small animals, including invertebrates, frogs and fish. There are at least 24 species in Australia (Wilson and Swan 2003), including the well known Snake-necked turtles (*Chelodina*) of south-east Australia, which are such popular pets. However, it is the 2 genera of similar shorter-necked turtles that are more common in the Australian tropical forest. Younger individuals have serrated edges to the shell that generally get smoother as they get older. Older females often develop large heads in a condition known as 'macrocephaly'. This feature may assist in feeding on hard shelled mussels; sometimes the head gets so big it cannot be retracted into the shell! (Wilson and Swan 2003). In the early morning these turtles are often seen basking on exposed rocks or logs to warm up, but they usually plop into the water at any sign of danger.

River turtles, Short-necked turtles
Emydura spp.

These turtles are usually medium-sized, with a smooth-edged shell up to about 30 cm. They have a short neck, and most distinctively, many individuals have a small yellow streak on the face behind the eye. There are around 6 species in Australia, with 1 species, **Krefft's River Turtle *Emydura krefftii***, being found in the creeks and lakes of the tropical rainforest, from southern to northern Queensland.

Saw-shelled Turtles
Elseya spp.

The species in this group are often known as 'snapping turtles', however, they are not related to the much larger true snapping turtles of the Americas, but only grow to a shell length of about 45 centimetres. They are better known as Saw-shelled turtles as the shells often have serrated edges, especially pronounced in juveniles.

As individuals age they often develop elaborate 'tubercles' along the neck and commonly develop 'macrocephaly'. They can sometimes be seen in the bigger fresh waterways, such as Lake Barrine. There are around 7 species in this genus in Australia, with 2 species being found in waterways in tropical rainforest: **Northern Snapping Turtle *Elseya dentata*** and **Saw-shelled Turtle *Elseya latisternum***.

Order Squamata
Geckos
Family Gekkonidae

Geckos are a family of mainly nocturnal lizards. They usually have slim bodies with long limbs, and often have fat or strangely shaped tails. Their skin appears rather soft, and includes a range of colours, from cryptic green and browns to striking bands. They usually have a flattened round head with a distinct snout. Their most distinctive features are the large lidless eyes which do not blink, but are licked clean by the tongue. Another feature is their ability to make noises, although most of the native species are quiet. In Australia, there are over 110 species. They seem to be more of the drier habitats, and are surprisingly less diverse in the tropical rainforest.

Leaf-tailed Geckos
Phyllurus spp. and *Saltuarius* spp.

These are large geckos with a leaf shaped tail and bark-like camouflage skin, found pressed up against tree trunks. Each species has a very restricted distributions in rain-forest along east coast of Australia. Pictured above is the largest; the **Northern Leaf Tailed Gecko *Saltuarius cornutus***, found in north-east Queensland.

Hemidactylus spp.

This genus is found throughout much of the world, but there is only 1 species in Australia; the introduced **South East Asian Gecko *Hemidactylus frenatus*** can be seen climbing around buildings in the rainforest (pictured left), rather than in the forest itself. Their loud call is a series of high pitched knocks; 'tchk-tchk-tchk'.

Carphodactylus spp.

This genus has only 1 species, which is endemic to Australia. Wislon and Swan (200) report the unique ability of this species to discard the tail, which then wriggles around and makes squeaking noises!

Chameleon Gecko *Carphodactylus laevis*. Relatively slender body, laterally compressed, long limbs, huge black eyes with 'black eye' smudge on snout. Carrot shaped tail; if the animal still has original tail, then it has several bold white stripes (pictured left, Hypipamee Crater). May be seen on ground at night in tropical rain-forest. Found only in North Queensland.

Dragons
Family Agamidae

Dragons are found from Europe, to Africa, Asia, Australia, and the south Pacific. There are some 70 species in Australia (Wilson and Swan 2003), but, like the geckos, they are more common in drier habitats and relatively less diverse in the Australian rainforest. Most dragons have large angular heads, sometimes with flaps, spines or frills. They have long bodies, strong hind legs and long toes, and very long tails. They generally have an upright posture and some species can run quickly on their hindlegs when alarmed.

Hypsilurus spp.

This genus is centred in the rainforests of New Guinea, with 2 species also found in the rainforests in eastern Australia. 'Rainforest' or 'Angle-headed' dragons are large lizards with big eyes and head crests. They are difficult to spot as they perch very still in the first few metres of a rainforest trunk. From here they watch for prey and avoid detection by predators; if spotted, they will slowly edge around the other side of the trunk. **Boyd's Forest Dragon *Hypsilurus boydii*** has white 'stones' embedded on the cheeks; it is found in north-east Queensland, and can sometimes be seen at Mossman Gorge (pictured right). The **Southern Angle-headed Dragon *Hypsilurus spinipes*** is found in the rainforest of southern Queensland/northern New South Wales.

Physignathus spp.

There are only 2 species in this genus; one is found in south-east Asia, while the other, the **Eastern Water Dragon *Physignathus lesueurii*** (pictured below, Daintree Eco Lodge) is found along the east coast of Australia. They have a large body, a fat head, a row of spines along the back, and a strong, laterally compressed tail which assists in swimming. As the name suggests, they spend much of their time around lakes or streams, and if threatened they plop into the water. They are common even in urban areas around southern Queensland, and in northern Queensland are best seen around Lake Eacham and Barrine.

Skinks
Family Scincidae

The skinks make up the largest of all lizard families, with over 1,400 species and continuing, with new species being discovered every year. They have adapted to almost every terrestrial niche on the planet. They are particularly diverse on the Australian continent, with over 370 species (Wilson Swan 2003) and are common in most habitats, including rainforest.

There is a huge range in size and shape in skinks, however, the typical garden variety; small, slim and fast, is indicative of the family in general. Most species have long bodies, thin limbs and very long tails. They tend to have large symmetrical shield scales on the head, and the body scales are usually small, smooth and round. As most species are primarily diurnal, they have small eyes. Being small means skinks are prey to many animals, but most have the ability of an efficient and familiar escape tactic. They have tail 'autotomy'; that is, they can shed their tail if they are caught or harassed, and are capable of growing another one back.

Calyptotis spp.

This small genus contains some 5 species of long skinks with comparatively tiny limbs, and a relatively long, thick tail. Their worm-like body shape reflects their habit of digging around underneath leaves and rocks in various habitats, including rainforest, along the east coast of Australia.

Lampropholis spp.

A genus of 11 species. These skinks are mainly brown or golden in colour and are found in wetter habitats along the east coast of Australia.

Glaphyromorphus spp.

This genus contains some 13 species of long skinks with comparatively tiny limbs, and a relatively long, thick tail. These rarely seen worm like lizards live mainly in wetter habitats in the Australian tropics, including rainforest.

Rainbow Skinks
Carlia spp.

This large genus of 30 species (Wilson and Swan 2003) includes some of the most commonly seen and well known skinks. They are found mainly along the east coast of Australia, especially in the north-east, and include many rainforest species. They are generally coloured in reds and browns and in breeding seasons, males often develop bright reddish patches to the head, throat or sides. They also have the habit of often making whip like signals with their long tails (pictured left, Josephine Falls).

Cyclodomorphus spp.

This genus is a group of 9 stocky blue-tongue like species. Only 1 species is found in the rainforest; the **Pink tongued Skink *Cyclodomorphus gerrardii,*** (left), ranging from Cairns south down to mid NSW. It is a large and strikingly banded lizard, which tends to be more active at night, and is capable of climbing vegetation with the aid of a prehensile tail. It feeds mainly on slugs and snails (Wilson and Swan 2003).

Egernia spp.

This genus has 30 species in Australia and includes some of the larger, more robust skinks. The main species found in the Australian tropical rainforest is the large and fat **Major Skink *Egernia frerei,*** sometimes seen near logs by the path at Mossman Gorge, but probably best seen on Fitzroy Island, and which also occurs in southern New Guinea (Wilson and Swan 2003).

Gnypetoscincus spp.

There is only 1 species in this genus; the **Prickly Forest Skink**, ***Gnypetoscincus queenslandiae*** is endemic to north-eastern Queensland, where it is restricted to the tropical rainforest. This unusual lizard has points to the scales along the back and thus appears 'prickly'. It is rarely seen as it feeds on invertebrates inside rotting logs (Wilson and Sawn 2003).

Shade Skinks
Saproscincus spp.

There are 10 species in this genus. Most species are a light reddish brown, often with gold, pale and dark flecking, sometimes a rusty line along the side of the tail, and a white spot at the rear base of the hind leg (Wilson and Swan 2003). They hide mainly in leaf litter on the moist forest floor along the east coast of Australia.

Water skinks
Eulamprus spp.

A group of about 15 medium sized species found in wetter habitats along the east cost of Australia. They tend to have copper or even golden backs, with black and white flecks or bars along the sides.

Monitors, Goannas
Family Varanidae

There are over 50 species of monitors and they are found in Africa, Asia and Australia, with the latter region having the highest diversity, with over half of all species. The name monitor originates in Africa, while the name goanna is commonly used in Australia, thus 'goanna' and 'monitor' refer to the same animals in Australia, and can be used interchangeably.

Monitors walk with their body held up, but still low to the ground, and with alternate movements of the opposite limbs. They have large feet and sharp claws that in many species are used for climbing and/or digging, and have a large, strong tail.

Monitors have a good sense of smell, using this to find their food, even if it is buried. Like snakes, they flick their forked tongues out to taste the air molecules. Most species are essentially predators, with prey varying from invertebrates to mammals to eggs, and they also scavenge. The predatory nature of goannas is reflected by the teeth, which are long, sharp and, in larger species, slightly serrated, and are also recurved for gripping prey (Steel 1997). Like many other lizards and snakes, they have fibrous tissue between bones that make up the skull, thus allowing a certain amount of flexibility, referred to as 'cranial kinesis' (Steel 1997). Monitors are not capable of masticating their food, but swallow in chunks.

Goannas are usually solitary, only seen in groups if brought together by a food source. It is difficult to tell male and females apart. The act of sex involves the male and female wrapping together closely. In tropical northern Australia this takes place in the wet season for terrestrial species (Greer 1989). Goannas usually lay their eggs in a hole dug in the ground if they are terrestrial species, or in hollow stumps if they are an arboreal species. Occasionally the latter types of goannas may also dig open a termite nest and lay their eggs inside.

Lace Monitor *Varanus varius* (pictured left). Large goanna. Much variation in colour and pattern (thus the scientific epithet *varius*), and often striking in appearance. Found all the way along the east coast of Australia, in a variety of habitats, including tropical rainforest. Often seen scavenging for food or lazing about at picnic and camping areas, such as at Cape Tribulation picnic area.

Snakes
Suborder Serpentes

Snakes probably evolved from lizards and into a very specialized group with a distinctive elongate form and no limbs. This form has obviously been very successful, as 2,700 different species (Shine 1998) swim, climb and burrow through a range of habitats, from the ocean through forests and grasslands, to deserts. Because of their wriggling locomotion and their often shiny appearance, many people think of snakes as slippery, however, their bodies are dry to the touch, and larger snakes are much like feeling a long muscular arm. They have some internal adaptations for their long body, including a long and thin liver, kidneys one after another rather than side by side, and a severely reduced left lung (Greer 1997).

Almost all snakes feed on other animals, with prey often being large and alive, and they have evolved various unusual features to accommodate these habits. Because snakes have a relatively thin body and are eating large animals, they have a very elastic head and body for engulfing their prey. The jaws can open very wide, and in many species the jaws literally separate from the skull to swallow the victim whole. Once inside, the lack of full encircling girdle and ribs enables the body to stretch considerably to accommodate the animal. To actually kill the prey so it doesn't tear the inside of the animal apart, snakes either constrict and suffocate their victims, or inject venom into them. This latter feature has resulted in the evolution of some of the most venomous animals on the planet. This probably accounts for much of the fear humans have of these creatures, however, they are usually reasonably shy animals, and most slide away when they hear humans thumping along a track. Despite the fears of many casual visitors, snakes are not encountered that often in the tropical rainforest. They are generally shy and blend in fairly well with their environment. Also fortunate is the fact that most of the more commonly seen species in the Australian tropical rainforest are not deadly to humans. Instead, it is groups such as the rear-fanged tree snakes (pictured below) and the non-venomous pythons that are observed in this habitat. The most dangerous snakes in tropical and subtropical Australia, region, such as Brown Snakes and Taipans, are rarely found in the rainforest itself, but rather in the adjacent savanna woodlands and sugar cane fields (so you are safer from snakes inside the forest than outside of it!). The best chances of observing snakes is probably on boat cruises, such as on the Daintree River and Lake Barrine.

Pythons
Family Pythonidae

The 25 species of Pythons are mainly found in the Old World tropics and subtropics, including Africa, Asia and Australia. They have reached their highest diversity in Australia, where about half of the world's species are found, and very well represented in the Australian tropical rainforest. Pythons are related to the Boas that are primarily in the American tropics, but differ from them in several reproductive aspects, such as the fact that they lay eggs. In contrast to most other snakes, female pythons show parental care; they wrap their body around their eggs and vibrate to keep them warm. Pythons generally tend to prey on vertebrates, and are notable for taking birds and mammals. They pick up chemical cues, and most can detect the heat from their prey with organs on their 'lips'. These sensor pits and their enlarged scales around the mouth (pictured below) are lacking only in a few species in the family, and are thus one of the most distinctive features with which to identify a snake as a python. They are one of the most commonly seen (and beautiful) groups of snakes in the tropical rainforest, and they include some of the major native predators in this ecosystem.

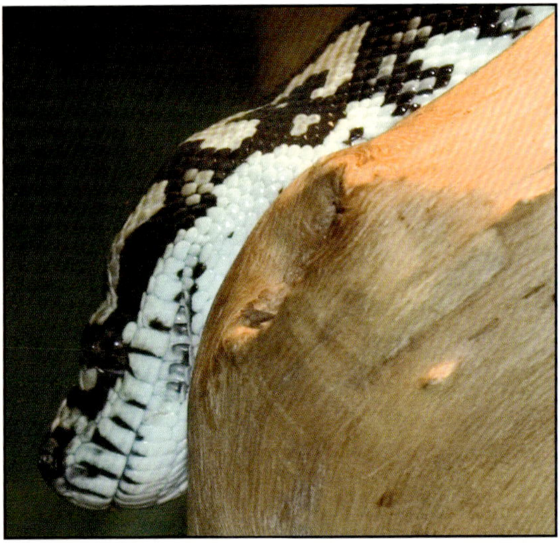

Antaresia spp.

This is a genus of 4 species of smaller pythons, include the smallest of all, the Pygmy Python of arid West Australia.

Spotted Python, *Antaresia maculosa.* Medium sized, grows to about a metre in length. Scientific name means spotted, and usually has dark spots and blotches on lighter brown background. Seem most common around rocky, hilly habitats, such as along Barron Gorge. Found in woodlands and forest from northern NSW to Cape York.

Morelia spp.

This genus of 5 species includes the largest and most striking of snakes seen in the tropical rainforest of Australia and New Guinea.

Carpet Python *Morelia spilota*. Widespread snake in Australia, and within its range appears to have some very different forms. In tropical rainforest of north-eastern Australia there is the 'jungle' version, a strikingly patterned form coloured yellow and black (pictured right, Rainforest Dome).

Green Python, *Morelia viridis*. Bright green. Grows to over a metre and a half. Famous in captivity for their equally bright yellow juveniles. Rests on branches in a distinctive posture (pictured left), reported to drink water collected in loops (Mirtschin and Davis 1992). Found in the rainforests and monsoon forests of north-eastern Cape York, and widespread in New Guinea.

Amethystine Python or Scrub Python *Morelia amethystina* (pictured below, Lake Barrine). Australia's biggest snake. Usually found in or near tropical rainforest. Strange, almost blue'ish sheen to scales in certain light. Generally arboreal (tree dwelling) and nocturnal, although can be seen sunning themselves during the day. Have been known to swallow bandicoots and rainforest wallabies. Sometimes seen on the Daintree River, but the best chance of seeing one is by taking a Lake Barrine cruise in the winter.

'Rear-fanged snakes'
Family Colubridae

This family is made up of 1600 species, although they are probably not a true group derived from one ancestor, but a collection of species that will eventually be separated into different families. Wherever there are snakes, these are the dominant group, except in Australia, where there are more of the generally more dangerous elapids, and the colubrids are limited to just 10 species, with only 1 endemic species. Here, they are found mainly in the tropics and along the east coast, but are not in the arid centre or the colder south. They are probably best represented in the tropical rainforest. The relative lack of species within Australia, their distribution in the more northern areas, and their close affinities with tropical Asian species, suggest a relatively recent invasion from south-east Asia. The colubrids are often called the 'rear-fanged snakes', as their fangs are further back in the jaw, or simply not present (Cogger 2000), and no species are considered fatal to humans in Australia. Because of lower toxicity, some species in this family in Australia constrict around their prey (Shine 1998).

Except for 1 species, all of the colubrids in Australia have a 'loreal' scale; this is the scale between the preocular scale (the scale in front of the eye) and the nasal scale (the scale with the nostril hole); the more dangerous elapids lack this scale.

Stegonotus spp.

There are about 11 species in this genus (Greer 1997), and they are found from South-east Asia, through New Guinea, to northern Australia. They are slight snakes, often glossy, with small dark eyes. The anal scale is single and the family loreal scale is present. They are mainly active at night, dawn and dusk. The diet varies, from mammals to reptiles and frogs, but also includes eating eggs of other snakes and lizards. Greer (1997) suggests the enlarged, broad teeth at the back of the top jaw are a specialization for slitting open eggs. They are also egg layers themselves. There are 2 species found in Australia; 1 is found mainly in New Guinea and within Australia is restricted to Torres Strait islands, while the other is found in the rainforest of North Queensland.

Slatey-grey snake, ***Stegonotus cucullatus***. Slender snake, grows to 2 metres (Mirtschin and Davis 1998), with flat head distinct from body. Dark brown, grey or almost black on top, but with a slight purplish tinge at certain angles (pictured left). Lighter underneath, including lower part of face. Bites readily if disturbed, but harmless (Wilson and Swan 2003). Found in various habitats, from woodland to rainforest, but usually not too far from water, including out-houses and water tanks (Burton 1998). Climbs well, but also semi-aquatic. Feeds mainly on frogs around creeks. Occurs in top end, most of Cape York and New Guinea (Burton 1998).

Boiga spp.

This genus includes about 25 species found from Africa, Asia, and northern Australia (Greer 1997), where there is 1 species.

Night Tiger, Brown Tree Snake *Boiga irregularis* (pictured below, Cape Tribulation road). Brown, usually with obvious 'tiger' stripes. Long thin body, may grow to 3 metres. Distinct round head, large 'dolls eyes' with vertical slit pupils. Large eyes because usually more active at night, when climbs through canopy feeding; young snakes take lizard eggs, adult snakes take bird eggs and sleeping birds (Greer 1997). Birds may be taken on the wing, and has been recorded attacking newborn pigs, puppies and babies (Greer 1997)! Has local reputation for biting when disturbed, mildly venomous, not fatal. Accidentally introduced to previously snake-free island of Guam after WW2, since then has decimated endemic birds and over-predated on lizards and bats (Greer 1997). Found in various habitats, including tropical woodland and tropical rainforest. Seen when travelling on ground between trees, frequently turns up in ceilings of buildings, such as sheds and outside toilets. Found from northern NSW to Indonesia.

Dendrelaphis spp.

There are some 16 species in this genus, and they are found through tropical Asia to northern Australia (Greer 1997). In Asia, they are often referred to as 'Bronze-backed Snakes'. However, they are best known as 'tree snakes', and are highly arboreal, with an individual of the Northern Tree snake recorded some 75 feet/23 metres high in the New Guinea canopy (Greer 1997). They are most active during the day, and are often noticed by humans when crossing the ground between trees. They are very slim snakes, with narrow heads. They mainly feed on small, fast, live prey such as lizards and frogs, and *D. punctulatus* has been recorded to eat the eggs and tadpoles of the cane toad without effect (Greer 1997).

The 2 species in the genus found in Australia are both found in the rainforest.

Common/Yellow-bellied Black/Green Tree Snake, *Dendrelaphis punctulatus*. Variable in colour; yellow, green, brown, black, blue (pictured right); thus, gets many different names. Found in range of habitats; woodland, gardens and forest, from Sydney, northern Australia, New Guinea. **Northern Tree Snake *Dendrelaphis calligastra***. Similar to previous species, but with less variation in colour, usually duller, black stripe on face, larger eye relative to head. Found in rainforest, from Townsville to New Guinea (Wilson and Swan 2003).

'Front-fanged snakes'
Family Elapidae

The elapids are often called the 'front fanged snakes'. The family includes the cobras, as well as various infamous Australian snakes such as King Browns and Taipans. Many members of this family are venomous and considered highly dangerous to humans. With over 80 species native to Australia, our continent is unusual in the fact that it has more of these venomous elapids than the generally harmless colubrids. However, while elapids are common in most habitats in Australia, from savanna to seashore, comparatively few of the more highly venomous species from this family are found deep within the tropical rainforest. Thus while many visitors are concerned about dangerous snakes in the rainforest, they really should be far more worried about the adjacent savanna, grasslands, sugar cane fields and roadsides!

To tell the difference between an elapid and a colubrid, have a close look at the scales on the face. The elapids lack a 'loreal' scale; this is the scale between the preocular scale (the scale in front of the eye) and the nasal scale (the scale with the nostril hole). Of course, it is recommended you do not have a close look at an elapids head while the snake is alive and in the wild! Only compare scales like this at close range if in captivity, or on a road kill specimen.

Pseudechis spp.

There are 6 species in this genus in Australia, of which only 1 is commonly found in the rainforest. They are all fairly large snakes, with fairly flat heads. Most are brownish or black in colour, and the genus is often referred to as the 'black snakes'. The anal scale is usually divided. They are all considered dangerous, and they will often flatten their neck out when threatened.

Red-bellied Black snake *Pseudechis porphyriacus*. Probably most common elapid snake seen in rainforest. Often just referred to as the 'Black Snake' in North Queensland, as many individuals seem to lack red on belly. In terms of toxicity and yield, is often placed about 10th in the world' (Shine 1998), however, while a bite from this snake is considered dangerous, is unlikely to result in death. Unfortunately, due to it's reputation as a venomous snake, they are often deliberately killed by Australians. Tend to feed on frogs in wetter areas.

Oxyuranus spp.

Coastal Taipan, *Oxyuranus scutellatus*. Handsome yellow to gold to brown snake, usually with distinct lighter colour head, 'evil eyebrows' and orange eyes. One of the more dangerous snakes in the world. Fortunately, not seen in the rainforest much, mainly on edge, and more common in adjacent sugar cane fields.

Furina spp.

These are 5 species in this genus. They are centred in eastern Australia and New Guinea, however, there is a species found in most parts of Australia, except for the cooler south-east. They are fairly small snakes, usually coloured in blacks and browns. There is often a contrasting patch of colour at the back of the slightly flattened head, and they have small dark eyes. They are terrestrial snakes, and usually hunt during the night, mainly eating sleeping skinks (Wilson and Swan 2003).

Brown-headed Snake, *Furina tristis* (pictured below). Small snake, usually grows to about 70 cm (Mirtschin and Davis 1992). Coloured dark brown with an almost purple/bluish tinge. Scale have light edges, resulting in a distinctive criss-cross pattern, especially on sides. Yellow patch on head. Feeds mainly on water skinks *Eulamprus* spp. Various authors report that snake is nervous and thrashes about when disturbed (Mirtschin and Davis 1992). Found in woodlands, but also in monsoon and rainforest. Range is from Daintree to Cape York, and onto southern New Guinea.

Crocodiles and Alligators
Order Crocodilia

Crocodiles are large and long, with relatively short limbs, thick and strong tails, and long snouts full of sharp conical teeth. They have existed in this general form for at least 200 million years. Despite their antiquity, they are often seen as one of the more advanced of the 'reptile' groups, including features such as 4-chambered heart (like birds and mammals, but unlike other reptiles) and careful maternal care. Fossil evidence shows that in previous times there have been large land based crocodilians; however, all the modern representatives are essentially aquatic animals. Reflecting this are adaptations such as high placed eyes and nostrils to enable seeing and breathing while submerged, and the strong laterally compressed tail which is used to powerfully propel the animal while swimming. Of the 3 families in the order, only the following is represented in Australia.

Family Crocodylidae

There are two species of crocodiles found in Australia; the smaller and much less dangerous 'Freshwater crocodile' and the larger and more dangerous 'Salt water Crocodile'. The former species is rarely associated with tropical rainforest, for it is normally found in smaller fresh water habitats up river from the estuarine and large freshwater habitats of the bigger species. It has been suggested that the freshwater crocodile only survives upriver from where the larger saltwater species can penetrate, due to the larger crocodilian predating upon the smaller. Following this reasoning, where the tropical rainforest is best developed - in north-east Queensland - the closeness of the ranges to the coast results in creek systems that are too short and fast flowing to allow significant upriver backwater habitat for the freshwater crocodile. In this region, the Saltwater Crocodile patrols the coastline and inhabits the mouths of rivers, and in large waterways such as the Daintree, it can be found well upstream in areas of tropical rainforest. Males may be looking for new territory, and smaller, younger individuals can occasionally be spotlighted in even some of the smaller creeks. Crocodiles can turn up in heavily visited areas, and as their predatory techniques are based on stealth, they may not be visible; thus it is wise to check the safety of any area before swimming.

Crocodylus spp.

Estuarine Crocodile, Indo-Pacific Crocodile, Salt Water Crocodile *Crocodylus porosus*. Biggest crocodilian in the world, also largest of all living reptiles. Not as abundant in waterways through rainforest as they are in mangroves and wetlands around the rest of northern tropical Australia. However, they can be found along sections of rainforest bordered rivers such as the Daintree, Mulgrave, Russell, Hull and even the Barron River through Cairns. Best spotted off boat cruises on these rivers, especially in the cooler months. Also known as 'Estuarine Crocodile' or 'Indo-Pacific Crocodile', the first name reflecting it's preferred habitat of estuaries, and the latter name reflecting it's large geographical range throughout tropical Asia and the Pacific.

Birds

Birds
Class Aves

There are almost 10,000 species of birds in the world (Perrins 2003). They are the most numerous of the terrestrial vertebrates, and in many ways the most successful, for they are found in basically every ecosystem all over the planet, from the desert to the polar regions. They are noticeable not just because they are diverse and widespread, but also because they are often colourful, noisy and mainly daytime active, therefore making them the most obvious animals in any particular region. In the Australian tropical rainforest where wildlife can be difficult to detect, birds are often the only relief from the monotony of green.

Most people would agree that the most distinctive features of birds are their feathers and their ability to fly. Fossils have been found of long extinct dinosaurs with feathers; however, birds are the only living animals with such a covering. These feathers provide much of the colour, and act as protection and insulation. And of course most species can do something humans have always fantasized about; they can fly.

Because they need to get airborne and stay in the air, birds are very lightweight animals for their size. Much about a their physiology and morphology is dictated by the animal's capability for flight. For example, their bones are full of many large air gaps, to reduce weight. This is also the reason why all modern birds lack teeth, for these are amongst the most solid of bones in other animal skeletons.

Like mammals, birds are endothermic (warm-blooded). This makes them fairly active, but also requires steady intake of food. Many species are very sociable animals, such as the colony inhabiting Metallic Starlings. They can also be very resourceful and intelligent, with some species showing what were once considered solely human traits, including tool use and artistic creativity, such as the bowerbirds and their bower creations.

All animals have developed their senses differently. Some species of birds have a good sense of smell, especially if they are carrion feeders such as the Brush Turkeys. However, it is sight where most birds so obviously excel. Many species, such as the predatory kingfishers, can detect the slightest movement. Many species have excellent colour vision. This colour perception is reflected in both the evolution of the bright and gaudy colours of many species of birds, as well as the colours of fruit aimed at species such as the fruit pigeons.

Despite the lack of obvious external ears, most species of birds have excellent hearing. This aural sense is reflected in the huge range of calls birds make. For example, in the Australian tropical rainforest, the calls range from the 'falling leaf' call of the gerygone, to the raucous screech of the cockatoo, to the sudden crack of the whipbird.

Birdwatching in the Australian tropical rainforest

For many people, birds and tropical rainforests are synonymous. And it is certainly no different in the case of the rainforests of Australia. The tropical rainforests of Australia are certainly one of the most exciting ecosystems for bird-watching in Australia, with many spectacular, rare and endemic species. In Australia, the tiny proportion of rainforest on the continent may be associated with up to half of the country's resident species. Bird-watching in the Atherton Tablelands in North-eastern Australia is easily some of the best on the continent. Most of the larger more obvious animals seen by visitors in the daytime are going to be birds, and the biggest native land animal in the tropical rainforest of Australasia is itself a bird.

However, the birds may not jump out at the more causal visitors. While keen birders find the forests teeming with new species, many more naïve visitors are disappointed. Many first timers to the rainforest, casual visitors and non-naturalists ask the question 'where are all the birds?' They are not always obvious to the novice, and finding them requires some patience and skill, as well as being able to deal with challenging conditions of the rainforest. However, some of the best birding can be done from the comfortable confines of verandas of rainforest lodges and hotels. Here, on the edge of the rainforest, the canopy effectively comes down to the ground and, as well as along roads and in clearings, provides great bird-watching opportunities. At wildlife and bird-watching lodges, where there is no hunting of wildlife and minimal disturbance, the birds become used to humans and even rare and elusive species become easier to see. Some of these accommodations will leave fruit out and have bird feeders to attract animals. Lodges include: Chambers Lodge, Possum Valley, Kingfisher Lodge and Cassowary House.

Ratites
Order Struthioniformes

This group includes mainly the large, flightless birds of the world. Today, they are generally only found in the southern hemisphere continents and countries. Africa has the Ostrich, South America has 2 species of Rhea (and the many Tinamous that are sometimes considered part of the order), New Zealand has 3 species of Kiwi, and Australia has 1 species of Emu and 1 Cassowary, and New Guinea has 2 more species of Cassowary. They have wings, but these are small and do not function for flight. The feathers do not interlock closely and thus hang loose and appear 'shaggy' like hair rather than feathers (Davies 1998).

Cassowaries
Family Casuariidae

Two species, the 'Single wattled Cassowary' and the 'Pygmy Cassowary' are restricted to the island of New Guinea, while the third species, the 'Southern Cassowary' is found in both New Guinea and Australia.

Cassowaries are large animals; in fact they are usually the largest native creatures within their rainforest habitat. Their body is covered in black hair-like feathers, with naked scaly legs that end in large feet with sharply clawed toes. In contrast, the head and neck have bright blue and red coloration. On top of the head is a unique structure, the casque. These birds are usually silent, but can emit a low, booming call.

Feeding ecology

Cassowaries eat mainly large colourful fruits, but are also recorded to eat carrion and other small animals. They are considered vital "keystone" species due to the fact that they eat, (and therefore distribute) so many large tropical rainforest fruits. In fact, the Southern Cassowary is probably the single most important animal seed disperser of the Australian tropical rainforest.

Studies have shown that the animal is an obligate frugivore; that is, the bird depends on fruits to survive (Stocker and Irvine 1993). So far, over 200 Australian tropical rainforest plants have been recorded to be eaten by cassowaries (Kroon and Westcott 2001). And even though the bird itself is hard to see, their scat can be quite common and conspicuous, and can contain up to 1 kilogram of seed (Stocker and Irvine 1983). One scat contained 13 seeds of about of about 6 centimetres in diameter (Stocker and Irvine 1983).

Life cycle

Cassowaries are basically solitary animals. Radio-telemetry studies have shown that cassowaries, especially in the uplands, have huge territories, sometimes of several kilometres, and that they may cover that distance in a day (Kroon and Westcott 2001). The only time cassowaries are seen together is when a father is with the chicks. The female may mate with several partners, then leaves to let the male look after the large green eggs. The male then raises the chicks, at which time he is known to be potentially aggressive.

Behaviour

The inner toe of the foot has a very large, sharp claw, probably used for defence. They have been known to attack people with these claws, and therefore should not be fed or approached. Thus, the Cassowary is considered one of Australia's few potentially dangerous large animals.

Southern Cassowary *Casuarius casuarius*. Very large, flightless bird; almost as tall as a person. Adults covered in black hair-like feathers. Neck is covered in bright red and blue skin, casque atop the head. Scat is large and often found in conspicuous place, (for example, on walking track), with lots of undigested seeds. Tracks are distinctive; large and three toed, with the middle toe much longer then the outside toes, distinguishing it from the Emu. Inhabits tropical rainforest, but also utilizes associated woodlands, mangroves, beaches and is known to steal fruit from gardens and farms. Restricted to north-east of Australia and New Guinea. Within rainforest habitat it is largest native animal.

Gamebirds
Order Galliformes

The members of this order are common all over the world, and are familiar to many people through the 'turkeys' (from North America) and the domesticated 'chicken' (derived from the Asian jungle fowl). The latter are occasionally seen wild in the Australian rainforest.

'Mound-builders'
Family Megapodiidae

There are 19 to 22 species of megapodes alive today (Watling and Talbot-Kelly 2001, Perrins 2003). These birds are found in Australia, Asia and the islands of the tropical south-west Pacific. There are 3 species in Australia; the famous Malleefowl inhabits the drier mallee country, but the other 2 species are found mainly in rainforest.

They are physically distinguished by their large feet, and this provides their scientific name of 'megapode', which means large foot. They use these big feet to claw and scratch in the forest floor for their invertebrate prey and other fallen food. Other than they look rather like chickens. However, they have an unusual nesting behaviour that sets them apart from all the other birds of the world.

Megapodes do not incubate their eggs by covering them with their body like most other birds. They kick together a mound of forest floor vegetation and let the heat from the resulting decomposition warm the eggs. Some species end up having the biggest nests of any birds in the world (Perrins *et al* 1996). They are also the quickest of all birds to fledge (Perrins *et al* 1996), for the precocious chicks are ready to run as soon as they have dug themselves out of the nest.

Brush Turkeys
Alectura spp.

Australian Brush Turkey *Alectura lathami*. Large, black turkey-like bird. Red head and yellow neck; these regions are featherless, giving it vulture-like appearance. Tail is unusual in that it spreads in vertical plane, and is used as signalling device to other birds and animals. Found on ground in forests along east coast; seem much more common in tropics. Can become tame in picnic areas and backyards where they scavenge off scraps, as they do at Mossman Gorge and many lodges.

Scrubfowl
Megapodius spp.

These megapodes are dark, chicken-sized birds that scratch around on the forest floor in lower altitude forests. They are normally shy and heard more than seen, but in some areas where there are people, they may become habituated, but still remain wary. There are about 14 species (Perrins 2003).

Orange-footed Scrubfowl *Megapodius reinwardt*. Chicken-like bird that scratches around in the rainforest litter with its large orange feet. Dark underparts and head, brown wings. Small head has crest. When disturbed, usually freezes, runs comically, or does a short flight into nearby tree. Usually much shier around humans than its relative Brush Turkey. Its calls include a loud double crow, often heard at night around tropical rainforest accommodations. Several pairs may work on nests that are contributed to over many years and end up becoming as a large as a car. Inhabits rainforests and monsoon forest in Australian tropics. Most common in lowland areas, around Mission Beach, Cairns, seen easily around the Cape Tribulation picnic area. Great example of huge mound can be seen along side of left side of Cape Tribulation road heading north opposite the Noah Beach camping area.

Order Pelicaniformes

This is a group of quite different looking birds including pelicans, cormorants and boobies. They are all water birds that feed on fish and they usually have webbed feet.

Anhingas
Family Anhingidae

Anhingas are tall dark birds, with large webbed feet and an upright stance, and due to their similarities with cormorants were once placed in the same family. However, they are distinguished from their relatives by their longer 'kinked' necks and their straighter, sharper bills, and are now placed in their own family. Like the cormorants, the Anhinga's wings are not waterproof and thus they must 'hang their wings out to dry'. The 2 species are found in many parts of the world, and subsequently have quite a few different names. The name 'Anhinga' is used mainly in the Americas, and was originally a Brazilian name for the bird. Other names are 'darter' and 'Snake bird'; the latter due to the way they swim with their body submerged, and only their long neck and head appearing out of the water.

Darter *Anhinga melanogaster (rufa)*. Considered an Australian subspecies of a wide ranging bird found in Australia, Asia and Africa. Straight, sharp bill is coloured yellow. Male is dark brown with chestnut coloured breast, female generally much lighter and greyish. Both have white stripe on face and neck behind eye. Call is a loud uneven rattle. Occasionally seen along waterways through the rainforest, such as the Daintree river and around Lake Barrine.

Cormorants
Family Phalacrocoracidae

Cormorants are usually black with some white or grey. They have a long body shape with short legs and big, webbed feet set back on the body so as they stand upright when on land. Unlike most waterbirds, the feathers of cormorants are not waterproof, which allows them to sneak into and slide through the water without disturbance. However, this also means they have to 'hang their wings out to dry' and thus they are often seen standing with their wings outstretched, standing upright on semi-submerged logs or boulders along streams. The following species can sometimes be found on the bigger waterways in the rainforest, such as Lake Barrine, and along the larger rivers, such as the Daintree.

Little Pied Cormorant *Phalacrocorax melanoleucos*. Small black and white cormorant with yellowish bill. Placement of eyes and markings on face give it a slightly angry appearance. Sometimes seen fishing along rainforest streams, often quite deep into forest. Their favourite perching rocks can be noted by the staining guano left behind.

Great Cormorant *Phalacrocorax carbo* (pictured below, middle of the log). World's largest cormorant. All black except for white cheek, yellow bill and (if you get really close) pretty blue eyes. Calls include various rattlings.

Little Black Cormorant *Phalacrocorax sulcirostris* (pictured below, both sides of the log). Smaller version of previous bird, but lacks colours on the face.

Raptors
Order Falconiiformes

This large and ecologically important order includes the eagles, falcons, kites, hawks and others. Due to their effective powers of flight, many of the species of raptors that are found locally are also reasonably widespread. Those found in the rainforest include a few species that are more specialized to live mainly in closed forests, but also several other species that are widespread, generalists found along large rivers and on the edge of the rainforest.

Crested or Pacific Baza *Aviceda subcristata* (pictured right). Smaller raptor with brown and white barring on breast. Only bird of prey in Australia with a crest. Sometimes spotted along roads through lowland coastal rainforest. Tends to stick close to canopy, feeding on small animals such as frogs and stick insects.

Black Kite or Fork-tailed Kite *Milvus migrans*, (pictured left). Medium sized, brownish with forked tail. Easily seen around more open areas outside of the rainforest, and occasionally inhabit edge of forest, fly above the canopy, or enter into gaps, such as above large bodies of water.

Grey Goshawk *Accipter novaehollandiae*, (pictured right). Handsome bird of prey, light grey upperparts, pale underneath, yellow-orange on legs and also behind beak. Large dark eyes. When seen quickly or from distance, appears distinctively quite pale. There is also a very white form. Found in various forests and occasionally woodlands, from Tasmania, up east coast and across northern Australia, and into New Guinea, Solomons and Indonesia (Pizzey and Knight 2003). May be relatively locally common raptor, yet secretive nature makes it elusive. Probably easiest to see on Atherton Tablelands, where occasionally spotted perched on wires in farmland on edge of rainforest.

White-bellied Sea Eagle ***Haliaeetus leucogaster*** (pictured right). Large white bird of prey with grey wings. In flight it's body and leading half of wings are white, while trailing edges of wings are black. Has relatively short, rounded tail. Often seen above coastal rainforest.

Cranes and Rails
Order Gruidae

Rails
Family Rallidae

Rails and the other members of this family are small to medium sized birds. Most species resemble chickens in shape and size, but are not related. They are often found around water and have large feet and long toes for distributing and evening out their weight so as they can delicately walk around and on top of aquatic vegetation. While some species, such as the Swamphens and Coots are easily habituated, most other species and shy and only occasionally glimpsed. They are found on all the inhabited continents of the world. Most live within grassy, reedy or swampy areas, but some are also seen on the edge of rainforests, such as the widespread **Purple Swamphen** (pictured left).

Buff-banded Rail ***Gallirallus philippensis*** (pictured below). One of the less secretive rails. Distinctive white eyebrow. Barring and small brownish smear on the breast. However, often simply appear plain dark as they are glimpsed running into cover on the edge of rainforest. Much easier to see on edge of rainforest on islands, such as Green Island.

Pigeons and Doves
Order Columbiformes

Pigeons and Doves have several distinctive features of their biology. They have a storage sac in the throat called a 'crop', which allows for digestion of quickly gulped food. They also have 'powder down'; instead of major oil glands to use for preening, these special feathers disintegrate to powder that clean and lubricate their main feathers (Crome 1998).

Different types of pigeons and doves

There is no true scientific or technical difference between 'pigeons' and 'doves'. Generally, the smaller pigeons often get called 'doves', but this is fairly inconsistent. However, the species in this order *can* be split into two different groups, but this is based more on their ecology and diet; there are those pigeons that feed on seeds, and those that feed on fruit. Those species that feed on seeds tend to live in more open, drier country, where there are grass and other seeds. This group includes some of the better known and more commonly seen pigeons, including most of the outback species.

However, the tropical rainforest has many more of the second type. The fruit eating pigeons that tend to that live in these wetter forests are lesser known, but more diverse, and certainly more striking. Although they tend to live high in the canopy and are shy and difficult to spot, they are worth looking out for, as they are mostly attractively coloured. And of course they are important to the ecology of the rainforest in their fruit eating and seed dispersing role.

Life cycle and behaviour

Pigeons often have a persistent courtship ritual which involves the male giving deep bows, lifting his tail and cooing loudly. In many species both parents help with the construction of the nest and the raising of the young. They also both produce a unique substance known as 'pigeon milk' in the crop, on which they feed their young (Crome 1998).

Geopelia spp.

The species in this genus are more dry country doves, but they can be found on the edge of the rainforest.

Bar-shouldered Dove *Geopilia humeralis*. Pigeon-sized 'dove'. Mainly grey with golden barring on the shoulder. Call is a distinctive and familiar 'kwook-a-woo', similar to the Peaceful Dove's doodle-oh. Found in New Guinea and Australia; in the latter they are distributed on the east coast and in the tropical north.

Peaceful Dove *Geopelia striata*. Cute little pigeon. Greyish upperparts, pinkish chest, barring over much of the body, and distinctive blue ring around eye. Call is a familiar and often heard 'doodle-do' and a rough, drawn out 'grrrrr..'. Often flushed out from the side of vegetation, having a distinctive undulating, almost flopping flight. Found from South East Asia, through New Guinea, to Australia and islands of the Pacific (where it is often known as the "Zebra Dove"). Within this distribution, can turn up in many different habitats, including edge of rainforest.

Lopholaimus spp.

This genus contains the **Top Knot Pigeon *Lopholaimus antarcticus***. This large, light and dark grey pigeon has 'Elvis Presley' swept back feathers atop the head. It is usually seen in small flocks flying above the forest canopy, and in other habitats, along almost the entire eastern part of Australia.

Columba spp.

This genus contains the ubiquitous Feral pigeon or Rock pigeon, common around urban areas.

White-headed Pigeon *Columba leucomela*. Largish black and white pigeon with red eyes, bill and legs. Occur in various habitats on the east coast, including subtropical and tropical rainforest, although not commonly noticed in the tropical rainforest around North Queensland, and better seen on the edge, and occasionally in towns such as Yungaburra.

Nutmeg Pigeons
Ducula spp.

This is a group of large, mainly black and white pigeons found around the islands of tropical Asia and the Pacific.

Torresian Imperial Pigeon, Pied Imperial Pigeon *Ducula bicolor* (pictured right). Large, almost fat, black and white pigeon. Seasonally common in the coastal rainforest of tropical Australia. Much of the population overwinters in New Guinea, with many returning to Australian coast to breed.

Chalcophaps spp.

Emerald Dove, Green Pigeon *Chalcophaps indica*. Beautifully purple-brown, bright green wings and red bill. Call deep, loud continuous 'cooo.....cooo'. Both sexes look after eggs; female at night, male during day (Schodde and Tidemann 1990). Usually found walking along ground; when disturbed may fly into canopy to perch, green blending in with foliage. Feed mainly on fallen fruit, native and weed; fig, raspberry, tobacco, lantana. Found in monsoon and rainforest in northern and eastern Australia. Best seen at Fitzroy Island (pictured lef). Also found New Guinea, through tropical Asia to India.

Leucosarcia spp.

Wonga Pigeon *Leucosarcia melanoleuca*. Large, chubby grey and white pigeon with black spots on underside. Found in various habitats, including rainforest, in eastern Australia, from Eungella in Queensland, to eastern Victoria. Best seen at O'Reillys and Royal National Park.

Macropygia spp.

Brown Pigeon, Cuckoo Dove *Macropygia amboinensis* (pictured below). From a distance a indistinct large brown pigeon. Close up, much more attractive and one can determine different shades of brown. Has long tail, thus the name 'cuckoo'. Sometimes seen feeding in pioneer species on edge of forest. Found along east coast of Australia, from Sydney, north to Cape York, New Guinea, and up into south-east Asia to Philippines.

Fruit Doves
Ptlinopus spp.

These 'Fruit Doves' are among the most colourful of all rainforest birds. They feed on fruit high in the forest canopy. Being so high up, they are rarely seen; however, their penetrating calls are one of the most distinctive sounds of the rainforest.

Wompoo Fruit-dove *Ptilinopus magnificus*. Biggest and most spectacular of the genus. Large, almost fat. Purple belly, bright green wings with yellow, light blue head, red eye and beak. Call is a deep, loud, umistakable rolling "woll-a-wak-a-woooooo" call. Feeds mainly in rainforest, but also found in monsoon and adjacent forests. Found on east coast from Sydney to Cape York, and onto New Guinea and surrounding islands.

Superb Fruit Dove, Purple-crowned Fruit Dove *Ptilinopus superbus*. Small green pigeon with bright purple crown and whitish belly. Male has orange red shoulders, a light purple throat and a large bluish black belly band. Call is a steady series of "hoo-hoo-hoo.." that does not accelerate. Found in various habitats, including tropical rainforest, along east coast of Australia, into New Guinea and Indonesia.

Rose-crowned Fruit-dove *Ptilinopus regina*. Small green pigeon with yellow or orange belly and, as common name suggests, rose crown. Call a familiar series of 'hoo-hooo-hoo's that accelerate. Found in range of habitats, including rainforest, along eastern and northern coats of Australia, up into New Guinea and Indonesia.

Parrots
Psittaciformes

'Parrots' belong to their own order, the Psittaciformes. They are mainly found in the southern continents and the equatorial regions of the world. There are many different kinds of parrots, yet they comprise a distinctive group of birds that most anybody would recognize because of a suite of familiar characteristics. The skin around the eye is often obvious, enlarged and appears fleshy, the bill is usually well developed, with a large hooked, down-curved upper mandible, over a smaller up-curved lower mandible. The arrangement of the feet is also unusual, with two toes pointing forward, and two backwards. Parrots also have some distinctive features of their behaviour and life cycle. Most species are monogamous, and many species appear to bond for life (Forshaw 1998). Most do not build nests, but raise their young in the existing hollows of trees, or excavate holes in banks, cliff faces, or termite mounds (Forshaw 1998). Parrots are not territorial and quite sociable, and can thus occur in large flocks. They are often nomadic, flying fast and wide to take advantage of varying resources. Most parrots have basic screeching calls that are not thought of as particularly melodious, but the highly sociable nature of parrots may have resulted in the development of many different calls for different purposes, including soft contact calls, warning calls, or calls made just previous to take off. There are more species in the neotropical rainforests than anywhere else, but it is in Australia where the group attains their greatest ecological, morphological and colour diversity, and they can be found from deserts to rainforests.

Cockatoos
Family Cacatuidae

Cockatoos are found mainly in Australia, New Guinea and surrounding islands. They are distinguished by usually having an erectile crest on top of the head and very heavy bills designed for cracking open hard fruits to feed on the seeds.

Sulphur-crested Cockatoo *Cacatua galerita*. Probably most familiar of all cockatoos. Large all white parrot with a yellow crest. Like many parrots, very sociable birds, seen travelling in huge flocks above rainforest canopy, and coming in at night to roost in large numbers in particular trees. While capable of sitting quietly in treetops, they are often noticed because they are such a loud bird, giving an ear-splitting screech as an alarm call. Occur in the south east, along the east coast, across northern tropical Australia, and into New Guinea.

Typical Parrots

These parrots are mainly green, with large and thick down-curved bills, and very short tails. In Australia, they are only represented by two essentially New Guinea species that also inhabit the forests of Cape York; the small Red Cheeked Parrot and the Eclectus.

Eclectus Parrot, *Eclectus roratus*. Large, short tailed parrot. Striking colours, males are bright green (pictured right, lower photograph, Sepik River, Papua New Guinea), while females are a pretty red and blue (pictured right, upper picture, Sepik River, Papua New Guinea). Usually noticed when flying high above canopy, more commonly male, who looks quite dark in silhouette. In Australia they are only found in Cape York, but are widespread in New Guinea, and also Solomons.

Fig Parrots
Micropsittinae

This is a group of small, dumpy parrots of tropical Asia, New Guinea and Australia, with 3 genera and 11 species (Juniper and Serpell 2003).

Double-eyed Fig Parrot *Cyclopsitta diophthalma*. Tiny and squat; Australia's smallest parrot. In flight, appear as small green round undulating rockets, with the blue in wings sometimes visible. When seen feeding in foliage, are a lovely green with yellow wash on wings and distinctive red patches above and below eye, resulting in common name of 'double eyed'. Call is very high pitched, vaguely metallic 'zeet! ..zeet!' when flying overhead above rainforest canopy, often heard around even urbanized areas, such as Cairns. However, once they stop to feed, usually quiet and difficult to see due to colour and size. Do indeed feed mainly on figs, but also eat other fruits and take nectar (Lindsey 1998). Unlike almost all other parrots in Australia that use existing holes in trees for nests, fig parrots actually excavate their own next, usually in a rotting trunk (Lindsey 1998). Found in tropical rainforest in 3 separate populations along east coast of Australia, also in New Guinea.

Lorikeets
Family Loriidae

The Lorikeets are small to medium sized parrots that feed primarily on the nectar of flowers. They have evolved finer beaks than the other parrots, and have unique brush-tipped tongues to mop up the nectar. Lorikeets are generally noisy and many species travel, feed, and roost in large groups. In Australia they feed mainly on the flowers of Eucalyptus and other Myrtaceae and they may travel far in search of their food. In the tropical rainforests, they are often seen at Evodias and Umbrella trees. They will also eat softer fruit if available. There are 2 species that can be seen in Australian tropical rainforests.

Rainbow Lorikeet *Trichoglossus haematodus*. As name suggests, literally all the colours of the rainbow. Found along the east coast and throughout the tropics of Australia where it can be very common and numerous. Also found in New Guinea. Often heard whizzing by in flocks above the rainforest canopy, but due to the height and thickness of the canopy, more commonly noticed on edge of forest or in urban areas where there are appropriate flowering trees.

Scaly Breasted Lorikeet *Trichoglossus chlorolepidotus*. Very pretty green parrot. The distinguishing feature that gives it common name are the yellow spots on green breast that appear as scales. Orange beak appears bright in contrast to greens of rest of bird. Often seen mixed in with flocks of Rainbow Lorikeets, of which it is similar in size and behaviour. Call is similar, but slightly more mechanical or 'tinny' in sound. Found in urban areas, woodlands and forest along east coast of Australia, from Sydney to Cape York.

Long-tailed Parrots
Family Polytelitidae

The species in this family are often considered more 'typical' parrots than the lorikeets or the cockatoos.

Australian King Parrot *Alisterus scapularis*. Beautiful green and red parrot. Very sexually dimorphic, with males having bright red head and breast, while female is more generally green. Usually travel in pairs or small groups of juveniles, the latter sometimes in a strange mix of confusing, undeveloped plumage colours. High 'ching ching' calls may be heard as they zoom overhead. Found in the rainforests and wetter sclerophyll forests along the east coast of Australia.

Rosellas, Broad-tailed Parrots
Family Platycercidae

This family consists mainly of the rosellas.

Rosellas
Platycercus spp.

The 'Rosellas' include some of Australia's best-known parrot species. The common name for these birds comes from an area west of Sydney known as 'Rose Hill'; the parrots were thus called 'rose-hillers', which then over time become 'rosellas'. The taxonomy of rosellas is a little confusing, with some species interbreeding in overlapping ranges, and many are known by various names that may not represent actual species. They are found in many habitats around much of Australia, missing only from the drier parts of the western and southern coasts, and much of the vast interior.

Crimson Rosella *Platycercus elegans*. One of prettiest Australian parrots, bright crimson, with blue on the tail, wings and 'cheek'. Like other rosellas, call includes a series of piping whistles and chattering; piping calls are three high, echoing whistles. Common in forests along the east coast, and often seen in adjacent urban environments. In the tropical rainforests of the north-east, they tend to be more common at higher altitudes.

Cuckoos and Coucals Order Cuculiformes

The 130 species of Cuckoo are found all over the world, including in the Australian tropical rainforest. They are more commonly heard than seen, especially when they arrive to breed in summer. The persistent, almost demented rising call of the Brush Cuckoo in summer can be heard around Cairns. The Coucals are often placed in a separate family, and are more of the Old World tropics. They include the Pheasant Coucal, often seen on the side of roads on the edge of the forest.

Owls Order Strigiformes

Owls are well known and instantly recognizable. They inhabit every continent except Antarctica. There are about 24 genera and over 160 species (Olsen 1998). They are well known for both their nocturnal and carnivorous habits. The owls are generally split into two different groups represented by two different families, and both groups have species in the Australian tropical rainforest.

The Monkey-face Owls *Tyto* spp. are the classic flat-faced owls. They are typified by the almost universally known and cosmopolitan **Barn Owl *Tyto alba*** (pictured right). They occur mainly in open areas on the edge of the forest, while the **Lesser Sooty Owl *Tyto multipunctata*** is endemic to the inside of the wet sclerophyll forest and rainforest of north east Queensland. It is a huge owl, especially when seen by day, with a distinctive scream that sounds like a "falling bomb'.

The Hawk-owls ***Ninox*** spp. are often known as the 'true owls'. They have small, rounded heads with curved beaks and staring yellow eyes. Their simple double-noted calls are often heard at night throughout Australia and this is sometimes the best way to track them, as they can be shy and cautious birds. In the tropical rainforest, this group includes species such as the ubiquitous Boobook Owl, as well as the rare Rufous Owl.

Nightjars, Frogmouths and Potoos
Order Caprimulgidae

This order contains various unusual night birds such as the nightjars and frogmouths. They are superficially similar in some ways to the owls, and are often mistaken for them; however, they are not closely related. They are nocturnal, have large eyes, often have bristles around the mouth and feed on various live prey.

Frogmouths
Podargus spp.

Frogmouths are bizarre looking nocturnal birds. They usually have brownish grey plumage, staring eyes, and large mouths. By day these birds may strike a stiff vertical stance on a branch that makes them appear as part of the tree in a remarkable act of camouflage. They are generally found in the Australasian region and up into tropical Asia. There are 3 species that can be found in the rainforests of Australia.

Tawny Frogmouth *Podargus strigoides*. Remarkable looking bird, brownish-grey with large orange eyes. Call is a distinctive low, deep, and continuous "ohm-ohm-ohm-ohm". While Papuan is more common in tropical rainforest of lowlands, Tawny is more common in rainforest of higher altitudes, such as around Atherton Tablelands and subtropical rainforest further south. Also extends into drier habitats over most of Australia.

Papuan Frogmouth *Podargus papuensis*. Larger than Tawny Frogmouth. Usually has redder eyes (in daylight or eyeshine) than the orange eyes of Tawny. Call, like Tawny Frogmouth, is strange, low continuous series of "ohm-ohm-ohm...". More common of the frogmouths in tropical rainforest of coastal lowlands, such as along Daintree River.

Order Coraciiformes

Kingfishers
Family Alcedinidae

There are some 86 species of kingfishers, with the highest diversity in New Guinea and surrounding islands (Fry 2003). Kingfishers are often divided into two main groups based on their feeding ecology. There are the 'true kingfishers' that do indeed fish, and these species tend to live near bodies of water such as streams, lakes and coastal areas. And then there are the 'land kingfishers', which include many larger species, such as the kookaburras. Naturally, there is much crossover between the two. All kingfishers are united by their diet of live prey, and this determines some of their physical features. They have a thick bill, strong neck and head and have a short squat body with a short tail. They watch their prey from atop an exposed perch and generally have excellent eyesight. Most species are coloured on the variations of blue and white, with some being amongst the prettiest of all birds. Kingfishers as a group are very widespread, being found almost all over the world except for the extreme polar regions (Forshaw and Kemp 1998).

Kookaburras
Dacelo spp.

The largest of the kingfishers in Australia are of course the Kookaburras. They have big heads and strong, thick bills that have evolved to catch larger, ground dwelling prey.

Laughing Kookaburra *Dacelo novaeguineae* (pictured right). Large kingfisher with big head. Brown and white with smatterings of blue on wing. Call is the familiar and famous rolling laugh. Found in eastern Australia in variety of habitats, including on edge of rainforest, such as along roadsides, or creeks and clearings within the rainforest. Sometimes seen on Bloomfield track or around Daintree village.

Alcedo spp.

Little Kingfisher *Alcedo pusilla*. Like a blue and white bullet; fast, bright, small, and hard to get a good look at.

Azure Kingfisher *Alcedo azurea* (pictured left). One of most beautiful of all kingfishers, coloured a bright blue with rufous underparts. Usually found near water, where it perches on tree branches or rocks waiting for movement. Often performs a peculiar movement, bobbing with its head and lifting its tail as it waits. In flight it can be very fast and direct, and may hover when hunting, plunging down for a quick splash as it grabs its prey.

Todiramphus spp.

Forest Kingfisher *Todiramphus macleayii* (pictured right). 'Neat' blue kingfisher with blue back and head, white underparts, and white patch in front of eyes. In good light, can appear striking bright blue, but in silhouette looks duller. More commonly seen in open habitats, such as farmland and woodland, but also on edge of rainforest, east coast and top end, to New Guinea.

Sacred Kingfisher *Todiramphus sanctus*. Similar to Forest Kingfisher, but duller blue upperparts and 'dirtier' underparts (pictured left). Found over much of Australia, New Guinea, Indonesia, also New Zealand, and many Pacific Islands.

Mangrove or Collared Kingfisher *Todiramphus chloris*. Similar to Sacred Kingfisher, but slightly larger, greener upperparts and clean underparts. Lives in mangroves, but also seen on edge of forests down river.

Paradise Kingfishers *Tanysiptera* spp.

This genus contains 6 species of kingfishers (Beehler *et al* 1986). They are colourful birds with elongated central tail feathers which make for very spectacular tails. They are found mainly in the forests of New Guinea, however 1 species migrates to Australia and breeds in the rainforest of north-east Queensland and Cape York.

Buff-breasted Paradise Kingfisher *Tanysiptera sylvia* (pictured right). One of our most magnificent birds; bright blue head and back, orange bill and, most distinctly, very long streamer-like, white tail. Arrive into coastal and lower altitude rainforest in Australia from New Guinea every year to dig nests in the small termite mounds on the floor of rainforest, and raise chicks inside. Regularly heard and seen at this time at Mossman Gorge, Jindalba boardwalk, and of course at Kingfisher Lodge.

Bee-eaters
Family Meropidae

'Bee-eaters' are small, fast, thin-beaked and often beautifully coloured birds found in Africa, Asia and Australia. Their common name is due to the fact that they swoop after insects such as wasps and bees, and actually remove the sting before gobbling them up. One species is commonly found in Australia.

Rainbow Bee-eater *Merops ornatus* This species is often heard and seen above the tropical rainforest canopy. It is a gorgeous bird of green, blue, rufous, and cream with black eye streaks and two small tail shafts. Their call is a distinctive rolling high trill, often made when flying. They head south to breed in the summer, and are thus more common in the tropical north in the dry season. At that time of the year they can be seen on the banks of the upper Daintree River, around the Cairns botanical gardens, and in various open areas where they may perch on dead trees or wires.

Order Passerines

This order is by far the biggest of bird groups, constituting about half of all known species. They are commonly known as 'perching birds' due to the arrangement of their toes, or 'songbirds', due to their sometimes elaborate territorial songs.

Family Pittidae

Pittas are found in the tropics of Africa, Asia and Australia. They are stocky but usually beautifully coloured birds. They are very elusive as they skulk around the ground of tropical forests, and with their loud calls are more commonly heard then seen. They feed mainly on snails, smashing them on a nearby 'anvil' rock, and thus leave evidence of broken shells. There are 4 species recorded in Australia, including the **Noisy Pitta**, ***Pitta versicolor*** of north-eastern Queensland, with a call usually described locally as a high pitched 'walk-to-work!'.

Australasian Treecreepers
Family Climacteridae

"Australasian treecreepers" are unique to Australia and New Guinea, and not related to treecreepers of other parts of the world. They are small birds with strong legs, relatively large feet, and de-curved bills. The well developed legs and feet are used to cling to the side of tree trunks, while the curved bill is used to pick at and underneath bark and rotting wood for invertebrates.

They have a very distinctive way of hunting; they make their way up the tree, usually spiralling up the trunk, then glide down to the next tree and start climbing up again. Unlike most other passerines that 'hop' with both feet in unison, treecreepers move their feet more independently, and thus cling to trees effectively, even walking upside down along branches.

White-throated Treecreeper *Cormobates leucophaeus*. Small bird, dark brown-grey on upperparts, streaks on the belly, a white throat, and distinctive orange dot on cheek if seen close enough. Call includes series of high piping whistles. Found in several different habitats along east coast of Australia. In tropical rainforest, seems common only at much higher altitudes; these populations look different from individuals in south-eastern Australia; generally smaller and darker, less white on throat, thus considered separate race (*minor*) known as the 'little treecreeper'.

Family Pardalotidae

This large and relatively recently erected family includes many different types of birds. In the Australian tropical rainforest they are represented by such groups as pardalotes, scrubwrens, and gerygones. These constitute many of the small and initially indistinct birds of the forest, also known colloquially as 'little brown jobs'. However, a bit of patience and closer inspection reveals many of the species to be quite attractive.

Gerygone spp.

'Gerygone' is pronounced 'Jerr-rig-a-knee'. This is now used as the common name, for they were once known as 'warblers', but are not related to the 'wood-warblers' of the Americas, nor the true warblers of the Old world. Telling them apart can be quite difficult; all species are small, with mainly brown or grey upperparts, pale underparts, red eyes and usually with a small white 'eyebrow' or a streak from the bill and going over the eye. They make up for the relatively dull plumage with their pretty calls, with many species creating a beautiful series of up and down notes. Their Greek derived name *gerygone* means 'born of sound', referring to these lovely calls (Cayley and Lindsey 2000). Many also build messy little nests that hang suspended off tiny threads, often built above rivers and resembling flood debris, thus also earning them another common name of 'flood-birds'. They tend to flit or hover around the foliage of trees, gleaning insects off leaves. There are 3 species that are associated with Australian tropical rainforests.

Brown Gerygone *Gerygone mouki*. Tiny plain bird. Red eye, white line above eye. Despite name 'brown', appear more of a grey. Usually flit around in the canopy in small groups, hovering and gleaning insects off leaves. Found in forests almost all the way along east coast of Australia. Quite common in rainforest, especially Atherton Tablelands. **Large-billed Gerygone *Gerygone magnirostris*** (pictured). Tiny brown-grey bird, red eyes, white eye ring. Found in woodland, mangroves and lowland rainforest of tropical coastal Australia and New Guinea. Bird and nest can be seen on Marrdja boardwalk, Daintree River. Similar Mangrove Gerygone can also be seen in mangroves and edge of forest in southern tropical rainforests. **Fairy Gerygone *Gerygone palpebrosa***. Prettiest and most distinctive of genus. Yellowish belly and a couple of white stripes on the face. Occasionally seen in lowland rainforest around Cape Tribulation.

Thornbills

Thornbills are found in different habitats all over Australia, especially drier, more open ecosystems. They are tiny birds that usually have darker brown back and wings, and lighter coloured underparts, often with a yellow wash, and with yellow or orange on the rump. Most species have streaking on the breast and on the face.

Acanthiza spp.

Mountain Thornbill, ***Acanthiza katherina*** (pictured right, Possum Valley). Tiny bird. Grey brown upperparts, lighter underneath with yellow wash. Black pupil, white iris, fine streaking around face. Call is a high pitched descending trill. Endemic to north-eastern Queensland rainforest, where it is found only at higher altitudes.

Pardalotus spp.

These are small but often brightly coloured birds. They hunt through the canopy (usually Eucalypts) for insects, mainly taking psyllid bug larvae and their lerps (the sugary excretion that forms the protective casing of the insect) off leaves. Of the 4 species of pardalote in Australia, only 1 is regularly found in tropical rainforest.

Striated Pardalote *Pardalotus striatus*. Small but striking. Cream coloured with bright butter yellow wash on belly, black on cap, face and wing, and big orange blob on forehead. Call is distinctive and often head; a persistent, sharp and far carrying morse-code like "pid-de-dip". Found over much of Australia in a range of different habitats.

Scrubwrens
Seicornis spp.

This group constitutes some of the smaller but still attractive birds of the rainforests of the east coast forests. They mostly have fairly dull plumage, but may have some distinct facial markings. They busily feed on insects and seeds, bouncing around the forest floor and in the first few metres of the vegetation. There are 6 species found in Australia, and all but 1 of those could be seen in the rainforest of the north-east. The best location to see these small birds is Possum Valley.

White-browed Scrubwren *Sericornis frontalis* (pictured right). Small. Brown wings and rump, grey head and neck, dull yellow-orange underparts, small white markings on wing. Black face with white stripes above and below bright yellow eyes. High pitched scolding whistling. Usually found in shrubs in a huge range of habitats in eastern and southern Australia, including rainforest and urban areas.

Fern Wren *Oreoscopus gutturalis*. Similar to previous species, but distinguished by much deeper chocolate brown colour. Found in higher altitude rainforest.

Yellow-throated Scrubwren *Sericornis citreogularis* (pictured left). Also similar to above, but with bright yellow wash to belly and throat (the specific name means 'lemon-throated'). Longer, pinker legs. Sweet warbling song, apparently imitates other birds (Cayley and Lindsey 2000) in southern parts of range. Found in forest and woodlands northern NSW and southern Queensland, with separate and distinct population in North Queensland, where they are found at higher altitudes.

Two very similar 'little brown jobs' are **Large-billed Scrubwren *Sericornis magnirostris*** (pictured below), and **Atherton Scrubwren *Sericornis keri***.

Large-billed seen in small busy groups flitting around vegetation, going higher than any other scrubwrens, calling "s-tchee, s-tchee", while Atherton hops quietly around forest floor with more subtle "chip, chip", usually by itself or in pairs. Eye of Large-billed tends to stand out from a plain face, and bill is larger and at certain angles appears almost to bend upwards. Atherton Scrubwren is endemic to mountain rainforests of North Queensland, and was not even recognized as a species until 1964 (Schodde and Tidemann 2000). Large billed found in rainforest all along east coast of Australia, and species formerly known as 'Tropical Scrubwren' in Cape York is now considered a race of this species.

Honeyeaters
Family Meliphagidae

The honeyeaters make up Australia's biggest family of birds, and thus there are many species found in the Australian tropical rainforest. Honeyeaters are, as the name suggests, nectar feeders, and thus vital pollinators for many Australian plants. They are in many ways the ecological equivalent of the African/Asian Sunbirds and the American Hummingbirds.

However, most honeyeaters are much bigger than the overseas nectar feeders, and most are not delicate enough to hover. Despite the name, honeyeaters do not exclusively feed on nectar, for while it is high in energy giving carbohydrates (i.e.; sugar); it is low in nutrients. Thus, most honeyeaters probably couldn't survive without including a few protein rich insects in their diet (Simpson *et al* 1996). Some ingest them as a consequence of feeding from flowers, but many 'honeyeaters' are actually more insectivorous than nectivorous.

Many of the physical features of honeyeaters are reflections of their nectar feeding lifestyle. Most have a bill that is curved to some extent, with Slater *et al* (1998) suggesting the bill shape largely reflects the types of flowers they are visiting. Another adaptation is the long tongue; it is split into 4 tiny hair like extensions at the end to give it the appearance of a brush, and this assists in soaking up the liquid nectar by capillary action. Some honeyeaters can move this tongue at more than 10 licks per second (Simpson *et al* 1996).

Generalizations of lifestyle and behaviour are hard to make with such a large and diverse family. Like many hummingbirds, honeyeaters can be territorial, often chasing others away from their nectar or insect source. Breeding seasons are extremely varied, reflecting the different times of nectar production in different plants (Egerton 1997).

A mated pair will usually construct a cup shaped nest suspended in the horizontal fork of a tree (Slater et al 2000). This may then be lined with softer material such as fur, feathers and plants. As with some other types of Australian birds, it is not unusual for assistant parenting from non-breeding pairs to occur (Egerton 1997).

Friarbirds
Philemon spp.

The 'Friarbirds' are among the largest, nosiest and, well let's face it, ugliest of the honeyeaters. They are fairly dull with shades of brown and grey, have black 'leather heads' and a knob on the top of the bill. They tend to be inhabitants of the woodlands rather than forest, but the following species is often found in the rainforest.

Helmeted Friarbird *Philemon buceroides*. Typical of the generally unattractive friarbirds, rendered even uglier by lack of feathers on back of neck, giving bird faint appearance of small vulture. Common in lowland jungle, as well as in neighbouring woodlands, mangroves, parks and gardens. Range of loud calls, including "wak-a-wear, wak-a-wear, wak-a-wear", and often heard around Mission Beach, Cape Tribulation or Cairns, especially in late afternoon.

Lichenostomus spp.

This is the biggest genus of honeyeaters, with about 18 species in Australia, and a few more in New Guinea. Several can be found in the mangrove, woodland and garden habitats on the edge of the rainforest, with 2 species being common residents of tropical rainforests.

Bridled Honeyeater *Lichenostomus frenatus.* Brown honeyeater with distinctive 'bridle' made up of yellow and black beak, white line to blue eyes, with white markings behind. Endemic to North Queensland, where usually found in higher altitude rainforests, thus once referred to as 'Mountain Honeyeater'. May scavenge around picnic areas, eg; car-park of Hypipamee Crater. There are also wintering populations in some coastal mangroves, such as at Marrdja boardwalk, where their antics and loud, rattling whistles can sometimes make bird very obvious.

Eungella honeyeater *Lichenostomus hindwoodi*, similar species, not recognized as separate species until 1978; inhabits higher rainforest around Eungella, central coastal Qld.

'Yellow Spotted Honeyeaters'
Meliphaga spp.

The *Meliphaga* are the type genus for the honeyeater family. There are about 10 species in the group, and they are centred in New Guinea (Beehler *et al* 1986). In Australia, there are 3 species and they are all among the most commonly seen or heard honeyeaters in the rainforest. They are all initially very similar in appearance, with several features in common; all have darker olive brown upperparts, lighter olive brown-yellow underparts, a thin yellow line along the face behind the bill, and a distinctive and large yellow spot on the face. In different species it varies from a half moon to a streak. However, this mark can vary with individuals and populations, and be difficult to get a good look at; thus habitat and call often need to be used to help identify the species.

Lewin's Honeyeater *Meliphaga lewinii* (right). One of most common and diagnostic birds of subtropical and upland tropical rainforests. Call is familiar and distinctive; fast 'machine gun' like series of whistles. Close up, tends to be slightly chubbier than other *Meliphaga*, and yellow patch tends to be fuller and more of a half moon. Will hang around picnic tables for hand-outs and in lodges in forest will often even come in and steal food. Can be aggressive to other birds. Common in rainforests along east coast of Australia, as well as gardens. In tropical north-east, is usually found at higher altitudes.

Yellow-spotted Honeyeater *Meliphaga notata* (right). Used to be called the "Lesser Lewins Honeyeater", as in many ways a slighter, warmer habitat version of that species. Has various calls, including some similar to the Lewins, but more muted; however, most distinctive call is a series of loud descending whistles. Bill is slightly longer and thinner than Lewins, and yellow patch is more of a rounded triangle. Usually found at lower altitude rainforests, as well as adjacent woodlands, mangroves and gardens in eastern Cape York, where can be very common, also found in New Guinea.

Graceful Honeyeater *Meliphaga gracilis*. Call a pleasant but subtle 'plik' call. Slighter bird than the other 2 species Usually found at lower altitude rainforest and adjacent woodlands in eastern Cape York and New Guinea. Often higher in the canopy, and less seen than other 2 species.

Phylidonyris spp.

There are 5 species in this genus in Australia. The genus name means 'pleasure-loving' (Cayley and Lindsey 2000)! Only the species below is found associated with tropical rainforest.

White-cheeked honeyeater, ***Phylidonyris nigra***. Black and white streaked bird with yellow on edge of wings. Large white patch underneath eye. There are 3 quite separate populations; south-west Australia, the southern east coast, and north-east Queensland. In latter region, can be seen on edge of forest in wet sclerophyll, usually around banksias at higher altitudes; can be seen outside windows at Tolga woodworks cafe.

Xanthotis spp.

The genus name refers to the 'yellow ear' (Cayley and Lindsey 2000). There are 3 species and they are found in the forests of New Guinea and Australia, with 2 species in the tropical rainforest of the latter region. They take nectar and fruit, but are mainly insectivorous.

Macleay's Honeyeater *Xanthotis macleayana*. Medium sized honeyeater with distinctive 'tartan' pattern, dark cap and orange patch around eye. Call a distinctive, fast and cheery "tsweet-swe-twseet!", best described by Pizzey *et al* (2003) as "a free tv! A free tv!". Endemic to tropical rainforest and some adjacent habitats of North Queensland. **Tawny-breasted Honeyeater *Xanthotis flaviventer*** is similar species, found northern Cape York and New Guinea.

Acanthorhycnhus spp.

Eastern Spinebill *Acanthorhycnhus tenuirostris*. Beautiful little bird, coloured rufous brown underneath, greyish on top, with black head, contrasting white bib, and very fine and sharp downward curving bill. Call is a series of piping whistles. Feeds regularly on nectar, and so often seen hovering at flowers with a distinctive whirring sound of wings, almost like a hummingbird. Found in the wetter environments along the east coast of Australia, in the tropics common only at higher altitudes..

Myzomela spp.

The birds in this genus are mainly small and delicate honeyeaters with thin, strongly decurved beaks. The males are quite often colourful, with reds being common, while the females are usually dull brown. This is the most widespread genus of honeyeaters, as they are found from Western Australia all the way across Australia into the central west Pacific.

Scarlet Honeyeater *Myzomela sanguinolenta*. Stunning and unmistakeable. Male has bright red head and chest and black wings (main picture right). Female and juveniles dull brown (smaller picture). Often found near blossoms of equally bright red Banksia and Callistemon. Found in various habitats, including rainforest along east coast of Australia, and into tropical Asia and islands of tropical Pacific. Can be seen in gardens at Lake Barrine.

Dusky Honeyeater *Myzomela obscura*. Subtly attractive honeyeater coloured chocolate brown, with slender, decurved bill. Found in various habitats, including rainforest, in tropical and eastern Australia, extending into tropical Asia. Probably best seen in gardens of Lake Barrine.

Australian 'Robins'
Family Eopsaltridae

The robins of Australia are not related to the English robins of which they were named after, but belong to their own family. They were so named because, like the English robins, the robins of the south-eastern Australian woodlands stood upright on a perch, flew out to feed on insects, and often had bright pink and red breasts. However, the 'Australian Robins' found in the rainforests tend to be more subdued in colour.

Grey headed Robin *Poecilodryas albispecularis.* Brown grey upperparts, buff underparts, grey head and white zig-zag marking on wing. Frequently hops up to alight a branch or perch at an upright angle on a vertical trunk. One of most obvious, distinctive, constant and ultimately annoying calls of the forest, a series of constant high whistles; 'whi.....whi.....whi..... whi.....whi.....whi...' and so on. Endemic to high altitude rainforest in the north east of Queensland. Within this restricted range, can be locally common. In some areas can become quite tame, e.g.; Chambers Lodge.

Eastern Yellow Robin *Eopsaltria australia*. Small plump bird with grey upperparts and often striking yellow underparts. Calls include a monotonous series of piping, and loud sudden "ch-yop ch-yop!". Found in a large variety of habitats in south-east and all the way along east coast of Australia. In the tropical north-east, tends to inhabit the wet sclerophyll forest on edge of rainforest.

Pale Yellow Robin *Tregallasia capito*. As name suggests, coloured a pale yellow on breast, with olive back, brownish-grey head, white throat and distinctive pale patch just above base of bill. Often seen on edge of clearings and tracks. Found in tropical and subtropical rainforests in two populations: northern NSW/southern Queensland, and north-east Queensland.

Family Pachycephalidae

The birds in this family are generally rather short, squat birds with a round head and a short strong bill. In fact, their family name means 'thickhead'. Some species are colourful, some are dull, but many are known, and named, for their calls. These calls are sometimes set off in reaction to a loud noise, like people yelling or a car door slamming. The Shrike-thrush (neither shrikes nor thrushes) have soem of the most familar and beautiful whistles of the Australian landscapes, based on a musical, loud and sudden 'pip-pip-pip-piphoooeee!'.

'Whistlers' are so named because of their fine musical whistles. The males are usually colourful, often with white bibs and darker heads, while the females are usually duller.

Golden Whistler *Pachycephala pectoralis*. Striking and attractive bird, the male having a bright yellow body, black face and white 'bib'. Found in a remarkably wide range of different habitats in Australia, from mallee to tropical rainforest, from south-west Australia to Cape York, Indonesia and widely found on islands throughout the tropical Pacific, with many subspecies. Tends to be more common at higher altitude in tropical rainforest of north east.

Logrunners
Family Orthonychidae

A small family endemic to Australia and New Guinea. One species, the Logrunner, occurs in the rainforest of south-eastern Queensland, as well as New Guinea, but is absent from the rainforest in between. The other species, the noisy Chowchilla, is only found in the rainforest of North-east Queensland, where it seems more common at higher altitudes.

Family Cinclosomatidae

This small family is endemic to Australia and New Guinea. One species, the Eastern Whipbird is found in tropical rainforests, more so at higher altitudes. It is also found in other habitats along the east coast.

'Old world Flycatchers' Family Dicruridae

Australia's native 'flycatchers' are part of a larger family that includes almost 140 species from Europe, through Africa, Asia and the western Pacific (Pizzey and Knight 2003). This family has been recently erected to also include a wide range of other old world tropical birds, such as the boatbills and the drongos. They are active little birds that usually sally out from a favoured perch to snap insects out of the air. Most species are small birds with relatively long tails. The beaks are usually short and, seen at the right angle, quite broad and flat. At a closer look, you may be able to see the relatively large and obvious stiff hairs around the beak; these are modified feathers called 'rictal bristles' and assist in catching their flying insect prey (Mead and Lundberg 2003). Most species build small cup nests in the branches of trees and bushes. Many of the species found in the rainforests of eastern Australia migrate in altitude or latitude in reaction to the seasons.

Monarch Flycatchers *Monarcha* spp. and *Arses* spp.

The Monarch Flycatchers are either coloured with grey upperparts and orange underparts (the *Monarcha* spp.) or black and white (the *Arses* spp.). They are active small birds, perching and darting through the understorey of the rainforest. Local populations fluctuate, as most Australian species undergo some migration in the southern winter, often to New Guinea in the southern winter.

Myiagra spp.

The flycatchers of the *Myiagra* genus are usually coloured cream underneath and gun metal blue grey above, with males tending to have rufous on the throat (pictured left). They usually inhabit mangroves, wetlands, woodlands and monsoon forest adjacent and grading into the rainforest.

Fantails
Rhipidura spp.

These little birds are usually paler underneath and darker on top, with black and white colours being common. They often have an 'eyebrow' of white on the face. Their name is derived from their long tail which they characteristically twirl around. This is believed to be used by the bird to flush out the insects upon which it preys.

Grey Fantail *Rhipidura fuliginosa*. Typical fan tail of the genus, dark grey with brownish belly, white throat and eyebrow. Common over much of Australia in huge range of habitats, and common in rainforest, especially at higher altitudes. Can be seen seasonally along walks through Mossman Gorge and at higher altitude, such as around Hypipamee Crater and Possum Valley. The populations in higher altitude rainforests tend to be darker individuals (pictured right).

Rufous Fantail has similar shape and behaviour to Grey Fantail, but has an attractive rufous tinge to the body. **Willie Wagtail *Rhipidura leucophrys*** (pictured left). Small black and white fantail. Familiar to most Australians through antics in urban areas over much of continent. Often observed on edge of rainforest.

Drongos
Dicrurus spp.

The Drongos once comprised their own family, but are now considered giant old world flycatchers. They are mainly a South-east Asian group of birds, but they also occur throughout the old world tropics, from Africa to south Pacific. The name comes from a native name for the bird in Madagasgar. Only 1 species is found in Australia, and it is common in tropical rainforest. The term 'drongo' for an idiot in Australia probably derives from a losing racehorse in the 1920's (Oxford 2004).

Spangled Drongo *Dicrurus bracteatus*. Large and glossy black with shiny red eyes and 'fish' tail. Has several calls, including whistles and scissor-like grating. Many individuals migrate to New Guinea in winter and return to lower altitude rainforest to breed in summer.

Cuckoo-shrikes
Family Campephagidae

Yet another example of birds in Australia being named for overseas birds that they supposedly resemble, but are not particularly closely related to; for cuckoo-shrikes are neither cuckoos nor shrikes. They are found throughout the Old World tropics and are called 'wing-shufflers' in Asia due to their habit of re-folding their wings when they alight. All fo the following species live in a iwde range of habitats, including tropical rainforest.

White-bellied Cuckoo Shrike *Coracina papuensis* (pictured right). Pale grey and cream bird with black eye stripe. Metallic trills can be heard in coastal lowland rainforest, such as around Mossman Gorge and Mission Beach, and also common around human habitations such as Cairns.

Black-faced Cuckoo Shrike *Coracina novaehollandiae*. Grey and cream bird with black face and throat. Also a common bird of woodlands, urban areas and farmlands adjacent to rainforest.

Barred Cuckoo Shrike *Coracina lineata*. Distinctive bird with obvious barring on belly and striking yellow eye. Metallic trills can be heard in coastal lowland rainforest. Tends to feed high in canopy and less obvious in open areas than other cuckoo-shrikes. Probably best seen at Mission Beach. Also seen in canopy of adjacent woodlands and mangroves.

Trillers
Lalage spp.

The Varied Triller *Lalage leucomela* is a medium sized black and white bird with a tinge of orange on the underside. As suggested by the common name, their call is a high pitched rolling trill. They are most commonly seen and heard at the edge of the rainforest canopy in lowland rainforest, especially on the coast and on islands, such as Dunk.

Australian 'Orioles'
Family Oriolidae

These Orioles are unrelated to the American Orioles, and are usually referred to as the 'Old World Orioles'. They are medium sized birds, with the males usually being brightly coloured, and often having combinations of yellow, green and black. Females tend to be much duller, often with streaking. Most species feed on fruit or insects. They are found from Africa, through tropical Asia, to Australia and the Pacific Islands.

Yellow Oriole ***Oriolus flavcinctus***. Medium-sized, yellow/olive bird with dark streaks, red bill and red eyes. Call a loud but pleasant, bubbling "yoh-yoh-yolll..!". Common in tropical rainforest along large rivers in Australia, such as Daintree and Mulgrave, and also around Cairns Botanical Gardens However, frequents top of canopy and can be difficult to see. One of best places is around pond at Daintree Eco Lodge. Similar but less striking Olive-backed Oriole common in adjacent woodland, similar call.

Figbird ***Sphecotheres viridis.*** Males in this species are a spectacular sight; olive green upperparts, bright yellow breast, and red 'spectacles'. Individuals in tropics tend to be brighter yellow than those further south. Females and juveniles much duller brown, with streaky underparts. As suggested by common name, they are indeed often found in fig trees. In quality rainforest they are hard to see as usually high in canopy; however, also found in urban areas including gardens and street trees, common in Cairns and Darwin, where they are easy to observe.

Family Artamidae

Many of the members in this new family were placed in different families until more recent genetic work placed them in the same group. The family is found mainly in the Australasian region, with some species stretching into tropical Asia. Most are variations on the colour scheme of black and white, and many Australians would think of them as typical 'bush' birds, with only a few found in the rainforest.

Butcherbirds
Cracticus spp.

Butcherbirds are predatory birds that feed on animals such as insects, lizards, snakes, small birds and small mammals, as well as scavenging Their common name derives from the habit of taking these items of prey and impaling it on, for example, a broken branch. The predatory habits are reflected in their strong straight bill, with a sharp hook at the end. Most species are generally found in woodlands, with only 1 species being found in rainforest. They readily adapt to scavenging around areas of human habitation.

Black Butcherbird *Cracticus quoyi*. All black, rather crow-like bird with thick silvery bill with black point. A predator in the rainforest, feeding on insects, lizards and even other birds such as sunbirds. Common in both mangroves and rainforest of lowlands, and often seen around lowland rainforest boardwalks, lodges and gardens, where they may scavenge from picnic areas (pictured right).

Currawongs
Strepera spp.

Currawongs are so named after their rolling calls. They are rather large birds, with thick, strong bills, and they often go after quite large prey, such as small mammals and other birds. They are also recognised as aggressive scavengers, common around picnic and camping grounds.

Pied Currawong *Strepera graculina*. Large, almost all black bird, with some white on wing that distinguish it from the other currawongs, and which is more obvious during flight. Eyes intense yellow. Live in wide range of habitats along east coast of Australia, including the higher altitude tropical rainforests.

Birds of Paradise
Family Paradisaeidae

There are some 43 species in this family. They are found mainly found on the island of New Guinea. In the better known polygamous species, the males are usually colourful and do elaborate display dances to entice the duller coloured females. There are 4 species also found in Australia; they are all found in rainforest along the east coast. One type of Manucode lives in northern Cape York.

Riflebird
Ptiloris spp.

This is the main genus of birds of paradise found in Australia. The loud single rasping cough-like calls or whistles may be heard more often than the actual bird is seen. The flight is quite loud, with 'steam train' sound wing beats. Both sexes are large birds with a long curved bill. Males are a spectacular glossy black with brilliant iridescent patches, on throat, head and rear. Juveniles and females are coloured in browns, with streaking or scalloping on underparts, and a line behind the eye; they look almost like giant honeyeaters, but behave like giant treecreepers. There are 3 species found in Australia.

Magnificent Riflebird *Ptiloris magnificus* is found tip of Cape York and New Guinea. **Victoria Riflebird *Ptiloris victoriae*** (pictured below, Chambers Lodge) is endemic to north-eastern Queensland. Most common at mid altitudes of Atherton Tablelands, but can also occur at sea level where rainforest is contiguous from high to lower altitude. Can be seen at feeders at Paluma, Chambers Lodge and Cassowary House. **Paradise Riflebird *Ptiloris paradiseus*** is endemic to south-eastern Queensland.

Bowerbirds and Catbirds
Family Ptilinorhychidae

The approximately 19 species of Bowerbirds and Catbirds are restricted to Australia and New Guinea. They are primarily rainforest birds, although a number of species have adapted to live in the drier tropical and central woodlands and grasslands of Australia and southern New Guinea. Bowerbirds are so named because of the bowers that the males build for courtship.

The males display near their structure to attract females. Their calls are usually quite distinct, with strange, often mechanical sounding clicking, buzzing and chirping. The males are usually more colourful compared with the duller females and juvenile males (young male Golden Bowerbird pictured at right), who may take years to gain their full plumage. The bowers themselves include a range of structures, from poles to avenues to simple cleared stages. Many are also decorated by collected objects, some species of bowerbirds favouring particular colours. Objects may be natural, such as leaves, flowers, berries, or unnatural, including balloons, pens, string, straws or condom wrappers.

Satin Bowerbird *Ptilinoryhnchus violaceus.* Male is beautiful but subtle sheeny purple, although may just look dark from a distance, and has purple eyes. Females and juveniles are duller green, often seen in small groups. Bower is two parallel walls about 30-40 cm high, decorated with many blue objects and occasional white or other colours. Bowers of tropical individuals seem smaller and less elaborately decorated. Call includes a series of mechanical burbles, like a robot farting. Found in rainforests along east coast of Australia. Most common in south east Queensland; best and most famously seen at O'Reilly's at Lamington, much scarcer in tropical north-east, where found only at higher altitudes, e.g.; Possum Valley, Mt. Lewis.

Tooth-billed Bowerbird *Scenopoeetes dentirostris*. Stocky, brownish bird with scaly underparts. Thick bill, with slight hook, which gives it common name. This bill used to cut off leaves, which the male then uses to decorate a cleared patch of display ground (pictured above), and thus earns it another common name 'stagemaker'. Restricted to higher altitude rainforests of north-eastern Australia.

Catbirds
Ailuroedus spp.

These catbirds are not related to the American 'Gray Catbird', which is in the same family as the mockingbirds (Sibley 2000). However, like that bird, they are named for their cat-like calls. They live in the same patch of forest year after year, and these loud wailing calls are for establishing this territory (Schodde and Tidemann 1990). Catbirds are coloured a very pretty bright leaf green on the upperparts, and have scaled breasts. They have thick, short bills and feed mainly on fruit in their rainforest habitat. Even though they are in the bowerbird family, they are not known to build bowers, and have only been recorded to occasionally pluck leaves and place them in a clearing (Schodde and Tidemann 1990). Unlike the showy bowerbirds, the catbirds form monogamous pairs.

There are 2 species found in Australia; the **Green Catbird *Ailuroedus crassirostris*** ranges from NSW to southern Queensland, while the **Spotted Catbird *Ailuroedus melanotis*** (pictured below) is found in north-east Queensland, Cape York and New Guinea.

Waxbills, including Grass-Finches
Family Estrildinae

The waxbills are an old world tropical family with over 150 species found from Africa, through tropical Asia, and throughout Australia and the western Pacific (Kemp and Evans 2003). The 19 species found in Australia are known locally as finches, but they are sometimes called Grass-finches to distinguish them from the many other sparrow like birds known as finches around the world. Waxbills are tiny birds with small conical bills. They have strong social structures, and thus are usually found in tight flocks, and often form life long bonds between male and female pairs. They are of course very popular as aviary pets, for many species have bright and contrasting markings. Being mainly grass seed eaters, they tend to be found in more open and grassy habitats and thus are not particularly well represented in the Australian rainforests. Even the species that do occur there tend to be found on the edges or in grassy clearings.

Notable species include that do inhabit the edge of the forest or clearings include the rare **Blue-faced Parrot-finch *Erythrura trichroa***, and the striking **Red-browed Fire-tail *Neochmia temporalis*** (pictured below).

There are several species that may be spotted on farms, grasslands and wetlands adjacent or merging into the rainforest, such as the **Chestnut-breasted Mannikin *Lonchura castaneothorax***, (right) and the striking **Crimson Finch *Neochmia temporalis*** (far right).

Sunbirds
Family Nectariniidae

Sunbirds are the African and Asian equivalent of the Australian honeyeaters and the American hummingbirds. There are over 130 species (Lack and Gill 2003), mostly found in Africa, but there are many in tropical Asia and 1 species has reached Australia.

Yellow-bellied Sunbird *Nectarinia jugularis*. Striking bird; bright yellow, male with iridescent blue-black patch on throat. Tiny and hummingbird-like, thin decurved bill, sometimes hovering in flight. More often found on edge of the rainforest rather than deep inside. Frequently builds nests around human habitation, such as along outside of lodge porches in Daintree, Mission Beach and around Cairns regions.

Flowerpeckers, Berrypickers
Family Dicaeidae

There are almost 60 species of these small, stubby birds. They generally feed on, as the common names suggest, nectar and small fleshy fruits. The 'Mistletoebird' is Australia's sole representative of this group.

Mistletoebird *Dicaeum hirundinaceum*. Male is black and white with bright red patches on chest and rump, while female is much duller. Have a fine curved bill and split tongue, family adaptations to feeding on nectar (Pizzey *et al* 2003). However, as suggested by common name, are better known for feeding on fruits of mistletoes. They shunt the small fleshy fruits through their digestive system in as little as 20 minutes. Seeds are often still very sticky, and may hang out in beads from bird's rear; upon landing, wipes bottom on branch to get rid of the sticky seeds. This disperses seed and allows the parasitic mistletoe to germinate in the canopy.

Swallows
Family Hirundinidae

This family includes the swallows and martins, of which there are some 15 genera and 89 species around the world (Sibley *et al* 2001). They resemble swifts in flight, but are not at all closely related. They are slightly less aerial than swifts, and come down to perch more often, and the flight itself is a little more regular than the fast and erratic dashing of the aptly named swifts. They are more common above the forest canopy, on the edge, and above water bodies within the forest, such as lakes (for example, Lake Eacham and Lake Barrine) or larger rivers (such as the Daintree River).

Welcome Swallow *Hirundo neoxena*. Tiny bird with dark bluish upperparts, greyish white belly, contrasting orange face and neck. In flight, wings sharp, V-shape in tail feathers, and dirty white belly and contrasting rusty neck often seen. Often observed flying above water, can regularly be seen perched on wires. In flight may emit single high 'chip' as swoop past. Small mud and grass nest, lined with fur and feathers, often attached under overhangs in caves and human-made structures. May nest in buildings, under bridges, on boats and river ferries (such as Daintree ferry, as pictured right). Found above various habitats over much of eastern, southern and central Australia.

White-eyes
Family Zosteropidae

This family contains the small yellow-green birds with white eye-rings found throughout the Old World tropics. They usually flit about the canopy in small groups. In Australia, they are more common in the lowlands, and seem to do especially well on islands. Three species can be seen in various habitats, including coastal rainforests; the **Pale White-eye *Zosterops citrinellus*** of eastern Cape York, the **Yellow White-eye *Zosterops luteus*** of the northern coast and tropical western coast, and the ubiquitous **Silver-eye *Zosterops lateralis*** found along the entire east and temperate coasts of Australia, as well as New Zealand (pictured left, Green Island).

Starlings and Mynahs
Family Sturnidae

The notorious 'European Starling' has been introduced into many other parts of the world, including southern Australia. However there is also 1 native species, of starling, and it is found mainly around tropical rainforest.

Metallic Starling *Aplonis metallica*. Smallish bird with glossy almost iridescent black plumage and bright, bulging red eyes (below, Dubuji). Arrive in huge numbers from New Guinea every summer to breed in colonies of nests suspended high in trees (right). Colonies change locations over the years. Often seen feeding nosily or swirling through and above canopy throughout lowland rainforest, as well as urban and agricultural areas.

Mynahs are tropical Asian birds that are best known for the ability of some species to mimic sounds, including human words. One species has been introduced into Australia. They are not to be confused with the Native Miners, which are a honeyeaters, and are usually found in Eucalyptus woodlands.

Common Mynah, Indian Mynah, *Acridotheres tristis.* One of most numerous and familiar birds in the tropics. Attractive brown birds with dark heads, yellow bills and yellow patch behind the eye. Easily seen bouncing around edges of road in farmed and urban areas on edge of rainforest.

Mammals

There are about 4,500 species of mammals and they are found in every habitat all around the world (Edgar 2000). Mammals are united in the one class by a combination of features. One of the most distinctive features is a covering of fur; even large mammals that appear naked have some reduced hair on the body. Bright colours are rare, with fur colour being drab greys and browns, as colour recognition is generally not a highly developed sense in mammals (except in many primates, such as ourselves). Instead, most mammals have highly developed senses of hearing and smell. These latter senses are important as many species of mammal (and certainly most species in Australia), are nocturnal. They are also 'warm-blooded' or endothermic; that is, they generate their own high and consistent temperature, usually have a high metabolism, and have an efficient four chambered heart. Mammal reproduction has some unique features. Young are usually born live, and mothers feed their young with milk from special glands. Variations on birth are reflected in the different types of mammals.

Types of mammals

The class of mammals is further divided into three subclasses; the Prototheria, the Metatheria and the Eutheria. Only in Australia and New Guinea are all 3 types of mammals present. The Prototheria are also known as the 'monotremes' and includes only the Platypus and Echidnas, and are found only in Australia and New Guinea. The Metatherians are better known as the Marsupialia. This group includes many members in the tropical Americas, however it is in Australia and surrounding islands where the marsupials have really evolved to such a spectacular effect and are currently such a dominant feature of the fauna. They include the familiar Koala and wombats, the kangaroos and possums, and the more unfamiliar bandicoots, carnivorous marsupials and Marsupial Mole. The rest of the mammals belong to the third sub-class, the Eutherians, better known as the 'placentals'. This group includes cows and sheep, rats and mice, cats and dogs, seals and whales, bats and monkeys and of course, ourselves. They are by far the most common and dominant mammals on the planet. In Australia, there are about 300 species of mammals. Many visitors and even residents of the continent assume that most Australian mammals are marsupials, however only about half of the continents native mammals are marsupial, with the rest being eutherian mammals, dominated by two orders; the rats and the bats.

Mammals in the Australian tropical rainforest

The tropical rainforest of Australia has representatives of all 3 main mammal groups; monotremes, (the platypus and Echidna), marsupials, (dasyurids, bandicoots, possums and kangaroos) and placentals (rodents and bats). Mammals are often difficult to see in the Australian rainforest, as most species are nocturnal, shy and small. However, there are a few mammals that you might be lucky enough to spot on a spotlighting tour, or wandering around your accommodation in the rainforest at night (padmelon pictured at Chambers).

Monotremes
Subclass Prototheria

Monotremes occur only in Australia and New Guinea. There are only 5 or so species; several types of echidna, and 1 species of platypus. The oldest remains of a monotreme have been found in South America, so prehistorically they have occurred, and perhaps even evolved, outside of the Australasian region (Strahan 1995). Both types of monotremes generally feed on invertebrates, and they can detect the electromagnetic impulses of their prey with their sensitive bills (Cronin and Westmacott 2000). Both the echidnas and the platypus lack real, developed teeth, and instead have horny pads in between which they grind their food. The males of monotremes develop spurs on the back leg, and in the platypus this becomes venomous in the breeding season. Monotremes are most famously distinguished from the other mammals by the fact that they lay eggs. Young are poorly developed after hatching and at a certain age are left behind in the burrow (Strahan 1995). These young monotremes are called 'puggles'.

Echidnas
Family Tachyglossidae

This family is comprised of about 4 species of echidnas in 2 genera, all found in New Guinea (Augee 2006). The Long-beaked Echidnas of the genus *Zaglossus* are only found on the island of New Guinea, and consist of at least 3 species, including recently discovered species (and including some further possible further splits). The Short-beaked Echidnas of the genus *Tachyglossus* includes 1 species that is also found across Australia. There are many different and interesting aspects to echidna biology and behaviour. They use their sharp digging claws to slash open nests, then use their long tongue to slurp up prey. Quite a bit of dirt is also ingested (Strahan 1998); this may assist in breaking down their ant, termite and other insect prey. The scientific name *Tachyglossus* means 'swift tongue', and it may slither in and out 100 times a minute (Strahan 1998). Their long snouts have been recorded being used as a snorkel when they are swimming! (Augee 2006). If disturbed or scared, echidnas bury themselves half into the ground, or wedge under a log or rock, so only the spines are visible. They are generally solitary animals (Strahan 1998), however, in the breeding season, receptive females may be followed by trail of up to ten males (Cronin and Westmacott 2000). Once mated, females develop a temporary pouch into which they lay a single egg (Egerton 1997).

Short-beaked Echidna, Spiny Anteater *Tachyglossus aculeatus*. Unmistakable. Size of small cat, covered with brown spines, long beak-like nose and tiny, squinty eyes. Colder climate individuals in the south tend to have more obvious hair (pictured left, Tasmania), tropical forms have less. Walks with distinctive rolling gait. Probably Australia's most widespread native mammal. Can be found in pretty much any terrestrial habitat, from desert dunes to tropical rainforest. Although widespread and common, not necessarily regularly or predictably observed, most likely due to its nocturnal habits, especially in tropical parts of continent.

Family Ornithorhynchidae

This family currently only contains the 1 species, the Platypus. They are one of the most unusual and famous of all animals. When reported to European scientists, they were thought to be a hoax, because they combined the fur of mammal, but the bill and webbed feet of a duck. They are basically a solitary animal, however, some may be seen in the same stretch of river where boundaries overlap, especially of more tolerant females. They live in a burrow dug into bank, often amongst tree roots, with a very tiny entrance just above the water line (Ryan and Burwell 2000). They spend much of their time foraging underwater for food. They have a layer of stiff guard hairs so that even after immersion the undercoat can remain relatively dry (Strahan 1998). They hold their breath for minute or so at a time as they gather aquatic invertebrates, such as crustaceans, and the occasional smaller frog or fish (Strahan 1998). The eyes are actually closed as they nuzzle around in bottom sediments with a bill that is sensitive to electromagnetic impulses sent out by prey. When enough food is in the cheek pouches, they rise to the surface to grind their catch between the hard pads, (rather than teeth), that are located in their jaws.

Platypus *Ornithorynchus anatinus*. Size of animal varies, but is always smaller than people expect. Those from colder climes, such as Tasmania, may weigh almost as twice as much as those found in warmer climes, such as North Queensland. Although probably more common that supposed, they are a shy animal, being more active at dawn and dusk than in the middle of the day. Can be relatively common in streams of tropical rainforest of Australia, but shyness mean they can be difficult to spot. Seem more common at higher altitudes, such as on Atherton Tablelands. Even though their status currently seems secure, could be seen as vulnerable due to fact that their habitat is the freshwater ecosystems east of the main dividing range (Strahan 1998); the very same habitat in which much of Australia's human population also lives.

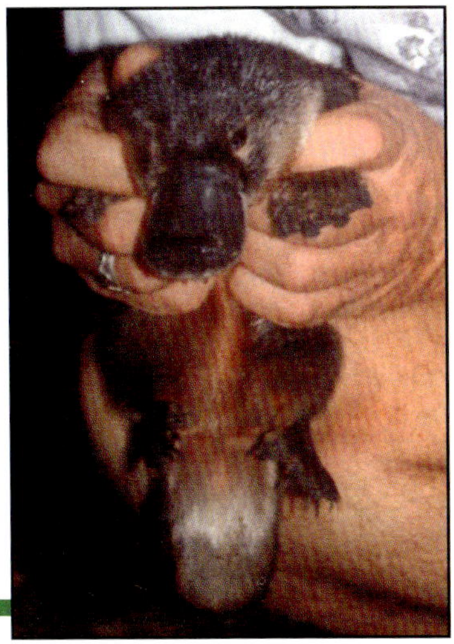

Marsupials
Subclass Metatheria
Order Dasyuromorpha

This order is dominated by one family which contains some 50 species of these generally predatory marsupials (Strahan 1998). This is the biggest family of marsupials, yet they are little known by the general public. This is probably the result of the fact that, unlike kangaroos and possums, the dasyurids are very rarely seen, for on the whole they are small, fast and generally nocturnal in habit. Being predators at the top of the food pyramid, they are naturally less abundant. Most dasyurids prey on insects and other invertebrates. However, they are well known for their tenacious nature and it is not uncommon for them to tackle prey larger than themselves, such as small birds and other mammals. The bigger dasyurids will also catch larger birds and feed on carcasses. In turn, they are prey for many Australian snakes, and it has even been suggested that the aggressive nature of dasyurids has resulted in the high incidence of strongly venomous snakes in Australia relative to the rest of the world (Martin 1993). Like many predators, they have evolved a fairly uniform morphology. Dasyurids tend to have four equal length limbs. The tail is not prehensile (Strahan and Cayley 1995) and may be covered in hair, be fluffy, or have a tuft at the end; in fact the groups name 'Dasyurid' means 'hairy tail' in Greek (Stahan and Cayley 1995). The head usually has large round ears and an elongated snout. Being predators, the long snout is filled with small, sharp teeth. Most people are only really aware of the two biggest species of dasyurid type marsupials, the 'Tasmanian Devil', and the now extinct 'Tasmanian Wolf', but most dasyurids are much smaller. Many of the smaller dasyurids have the cute appearance of small rodents, and were thus often described as 'marsupial mice'; they are now known under various genus names, such as *Antechinus*.

The main species likely to be seen in the tropical rainforest is the **Yellow-footed Antechinus *Antechinus flavipes***, but even then usually only as a tiny, almost imperceptible brown flash on the rainforest floor. These relatively rarely seen marsupials are often confused with the much more common *Melomys*, a group of native rodents. However, Antechinus can be distinguished by their splayed out feet, sharper (rather than rounded) snout, and double lobed (rather than classic round) ears. Others dasyurids vaguely recalled pole-cats, and were thus named 'native 'cats'; these are now known as 'Quolls' (Strahan 1998). The largest species in Australia, the **Spotted-tailed Quoll** or **Tiger Quoll *Dasyurus maculatus***, is found in the tropical rainforest, but is very rarely seen.

Above: The Spotted-tailed Quoll or Tiger Quoll

(Australian) Bandicoots
Order Peramelemorphia

Bandicoots are a distinctive order of only about 19 marsupials in Australia and New Guinea (Egerton *et al* 1997). The name 'bandicoot' is not a particularly apt one. It is of Indian, not Aboriginal origin, being named after large rodents found in Asia that have the common name of 'bandicoot' and the scientific name of *Bandicota* (Prater and Barruel 1993). They have also been compared to 'badgers', for the derivation of the scientific name means 'pouched badger' in Greek (Strahan and Cayley 1997). Bandicoots are roughly rabbit-sized and shaped. They have shorter, but strong, forelimbs for digging (Strahan and Cayley 1995) and longer back legs used in their bounding, rabbit-like gait. They are not grazers like rabbits however, but are omnivorous and have a more shrew-like, pointed snout. In fact, while they probably most physically resemble rabbits, some authors suggest they probably best resemble ground shrews or even small pigs in their ecology. Ultimately, Australian bandicoots are marsupials and so obviously not at all related to any of these other animals.

Most species spend much of their feeding time digging into the soil with their sharp claws and pushing that long snout in after invertebrates upon which it feeds. They are also noted to feed on fruits, tubers and small mammals (Strahan 1995). Being marsupials, bandicoots give birth to fairly undeveloped young. Some of the bandicoots are noted for having amongst the shortest gestation period of any mammal known, at just under two weeks (Strahan 1995). These undeveloped young then continue to grow in a pouch which in some species may not be much more than a flap of skin. With such a fast reproductive cycle, if conditions are good, some species of bandicoot may breed and raise offspring every seven weeks (Egerton *et al* 1997).

Northern Brown Bandicoot *Isoodon macrourus*. Size of large rabbit, with general pear-shape body. Ears are much shorter than long nose bandicoot. Covered in stiff, spiky brown hair. Long nose covered with fur. Usually the more commonly encountered bandicoot in rainforest because occurs on edge, along roads and in more open habitats.

Long-nosed Bandicoot *Perameles nasuta* (pictured right, Black Mountain road, Kuranda). Mainly grey with brownish tinge. As name suggests, has long shrew-like nose, noticeably longer than brown bandicoot and with some fleshy pink visible underneath sparser fur. Tail also has sparser hair and the ears are much longer. Is a cute critter that tends to squeak like a tight balloon and has a quick bounce when alarmed. In North Queensland, is usually found a bit deeper in forest than the brown bandicoot.

Order Diprotodontia

This large order contains the mainly 'herbivorous marsupials', including the Koala, the wombats, the kangaroos and various families of possums.

Possums (various families)

The name 'possum' is a fairly vague and not very scientific term. Other than both being marsupials, the Australian/New Guinea 'Possums' are not closely related to the American/Neotropical 'Opossums'. And in fact, within Australia, the term 'Possum' actually covers a few quite different families of animals.

There are a few characteristics that unite the different Australasian possums, although these are often consequences of ecology than a reflection of relatedness. All species are mainly active at night, thus their senses of smell and hearing are highly evolved. Many species have scent glands, and individuals may be capable of recognizing different characteristics, such as sex and age of another individual, by scent alone. Many species have large ears and react to seemingly small sounds in the forest. Their sight is sensitive to light, and all species have varying degrees of eyeshine from flashlight reflection. Different species have different colour eyeshine, and in fact this aids in identifying different species in the Australian tropical rainforest. All possums are quite arboreal, and many have a strong, usually furry tail with a naked patch of skin underneath which assists with grip (Egerton 1996). Many species have also evolved feet that are effective grippers of branches by having a larger, single opposable digit.

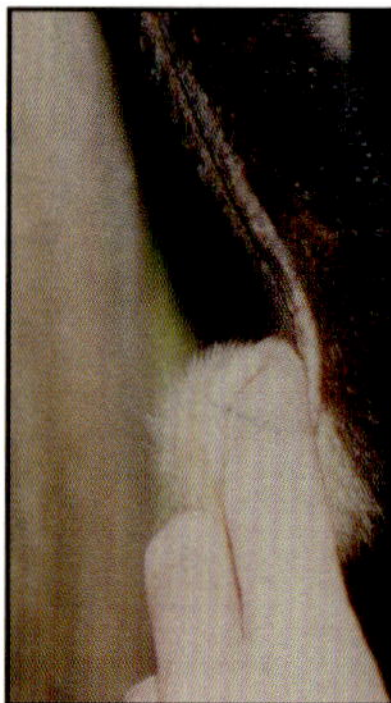

Above: the prehensile tail of a possum

Below: the opposable digits of possum feet

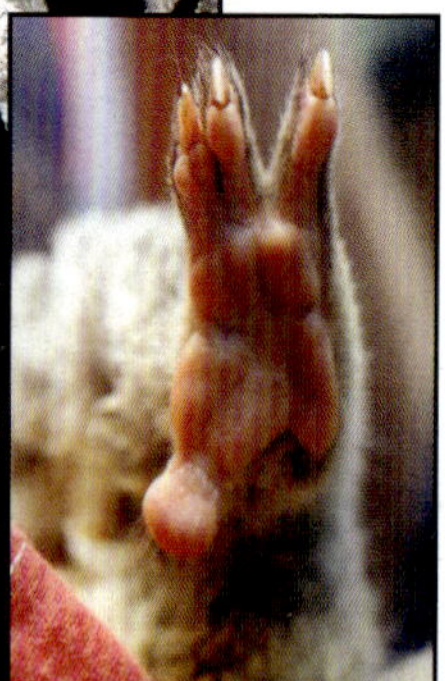

Wrist-winged Gliders
Family Petauridae

Gliders are usually grey or brown in colour, with a bushy tail and a darkish stripe on the back that often extends to the forehead (Strahan 1998). They are referred to as the 'wrist–winged gliders', as most have a loose flap of skin between the wrist and ankle on the same side. This appears as a loose fold of skin when climbing, but when jumping from tree to tree the skin is stretched and acts as a gliding surface. They have been reported to glide up to 100 meters (Egerton 1997), although in the thicker tropical rain-forest these glides are usually more like short jumps, the gliding itself is rather difficult to detect at such speed. Gliders use their sharp lower incisors to chew into trees and then lick the resulting sap and gum (Egerton 1997). This is especially so in winter (Strahan 1998) when their other foods such as insects and other invertebrates are rare. They also nibble and suck on fruit and flower nectar. Being nocturnal, their sense of small is important. In the more sociable species, they use scent trails to identify other individuals of the same species (Strahan 1998).

Sugar Glider *Petaurus breiceps* (pictured below at Chambers Lodge). Not often seen, but can be common if enough tree hollows for them (Strahan 1998). Occasionally heard; call is a high pitched, even, fast 'yip-yip-yip....", which serves as alarm call, indicating there is potential predator in area, such as a cat or owl (Strahan 1998). Most live in small family groups, especially when it is cold and they can huddle together to stay warm (Strahan 1998). Younger members of group are usually evicted before 12 months of age, and subsequently there is a high rate of mortality for young gliders (Egerton 1997). As with other gliders, scent and smell are important in social interactions. Recent research indicates secretions from dominant males can suppress reproductive activity in other males in the group (Egerton 1997). Widespread, being found in range of habitats, including rainforest, in the south east, along the east coast and throughout the top end of Australia and into New Guinea (Egerton 1997). In the tropical rainforest of north Queensland, best seen at Chambers Rainforest Apartments.

Family Phalangeridae

This family includes the brushtail possums, the various types of cuscus and the Scaly-tailed possums, with a total of some 24 species (Flannery 1995b). They are in fact the most widespread of all Australasian marsupial families, from brushtails in Tasmania, to cuscus in the northern Sulawesi Islands, Indonesia.

Cuscus

The cuscus are 2 genera of possums. They usually have much smaller ears than their brushtail relatives. They are found mainly in New Guinea, but several species also live in the surrounding islands of the south-west Pacific, and there are 2 species found in the forests of the Cape York region of Australia.

Common Spotted Cuscus *Spilocuscus maculatus*. Distinctive, strange looking possum. Adult males usually pale, with blotches, females and juveniles usually greyish. Very short ears barely visible, giving appearance of being 'bald'. Found in New Guinea and some surrounding islands, and also found in Australia (Flannery 1995a, Flannery 1995b), where they are found in various forest types in northern Cape York.

Southern Common Cuscus *Phalanger intercastellanus* (pictured below, Papua New Guinea). Small possum, greyish brown above, paler below, usually with obvious dark striped down centre of head. Bulging yellowish eyes. Compared to spotted cuscus, this species has medium sized ears, and obvious snout. They have been recorded to feed on a wide variety of plant materials, including leaves, flowers (from *Carallia*), fruits (from Red Cedar) and seeds (from Black Bean) (Flannery 1995a). Found in southern New Guinea and surrounding islands, and also in Australia. Here, they are restricted to rainforests of eastern Cape York. Fairly sparse in habitat, but can be seen around villages in southern New Guinea lowlands, both feeding in gardens and kept in cages.

Brushtail Possums
Trichosurus spp.

Brushtail Possums are quite omnivorous; they'll eat leaves, fruit, flowers and french fries. They are generally the only type of possum likely to be seen on the ground, where they may take handouts from people and scavenge around human habitations. The 'Brushtails' are probably the most vocal of the generally quiet marsupials, and their large fox-like ears reflect how important vocal communication is. Scent is also used, with a musky smelling substance produced from glands near the anus and on the chest; thus 'brushies' tend have a particular odour. They are found in a wide variety of habitats throughout much of Australia, including the tropical rainforest.

Common Brushtail Possum T*richosurus vulpecula*. Can be found, in one form or another, throughout much of Australia. Most vocal of all Australian marsupials, with growls, chattering clicks and screams (quite eerie and scary in the bush at night). Tail is quite prehensile; even a heavy, well-feed, urban possum can hang upside down off a branch by just its tail. Rather large, solidly built possum with dark furry tail. Colour of fur ranges quite a bit, from black form, to more common grey form, to a 'coppery' rainforest form. **Coppery Brushtail Possum T*richosurus vulpecula johnstonii*.** A sub-species (that may eventually be classified as separate species). Like many other possums of Australian tropical rainforest, is generally found at higher altitudes, such as in Atherton Tablelands. Can usually be seen when spotlighting at Hypipamee Crater, often even around carpark.

Above: Coppery Brushtail Possum

Ringtail Possums
Family Pseudocheiridae

This family includes at least 15 species in a range of genera (Flannery 1995a). They are found in New Guinea, Australia, and some surrounding islands, with 7 species in Australia. They are not generally quite as wide ranging in their diet as other possums, such as the gliders and brushtails. They may feed on fruits and flowers, but generally the bulk of their diet is leaves (Egerton 1997). This results in 'ringtails' being more sluggish and appearing rather 'dopey', although many are capable of energetic leaps within the canopy. However, they are excellent climbers with strongly prehensile tails (Strahan and Cayley 1995). They are also usually quieter than many other possums; their less reliance on vocalization in communication is reflected in their smaller ears.

Ringtails are found in a variety of habitats in Australasia and surrounding islands, although they have a tendency to be more of a closed forest possum than the 'gliding possums' and the occasionally terrestrial 'brushtails'. They have speciated most in the higher altitude rainforest of New Guinea and north-east Queensland.

Pseudochirulus spp.

The species in this genus are found mainly in New Guinea and surrounding islands. There are 2 species found in Australia.

Herbert River Ringtail *Pseudochirulus cinereus*. One of most beautiful Australian possums, coloured dark brown on top and cream on underparts, giving appearance of almost black and white 'panda'. Symbol of Queensland National Parks and Wildlife Service, although most Queenslanders would not have seen, (or even heard) of this species. Restricted to tropical rainforest in North Queensland above 350 metres or so (Menkhorst and Knight 2002). Around higher altitude rainforest in Daintree region is much plainer **Daintree Ringtail Possum *Pseudochirulus cinereus.***

Hemibelideus spp.

This genus only contains 1 species. It is restricted to the higher altitude rainforest of North-east Queensland.

Lemuroid Ringtail Possum *Hemibelideus lemuroides* (pictured at right, Hypipamee Crater). Usually a fairly dark possum with little in the way of distinguishing colours or markings, being greyish brown with thick, furry tail. Most distinctive feature is probably bright yellow eye-shine. It is only rainforest possum of size regularly found in groups. Restricted to upland tropical rainforest in north-east Queensland, where they often seen very high up in canopy. In right habitat, can be locally common.

Pseudochirops spp.

This genus contains 4 species found in New Guinea, and 1 species restricted to Australia.

Green Ringtail Possum *Pseudochirops archeri* (pictured below at Possum Valley). One of few naturally green furred mammals in world, though it tends to be more of an olive than green. Very distinctive pair of cream patches underneath eyes, and another pair underneath ears, as well as dark and lighter 'racing' stripes down back. Unusual as does not build a drey to roost in, but sleeps out in open during day, greenish fur making animal hard to spot in canopy. Like most of other possums in Australian tropical rainforest, generally restricted to the higher altitudes, although this species is also common at mid altitudes around Yungaburra, and is occasionally reported in the lowlands. Found only in North-east Queensland.

Kangaroos and Wallabies
Superfamily Macropodoidea

This is a diverse but recognizable group of marsupials. There are at least 65 species of 'kangaroos', with most being found in Australia and also New Guinea. Despite the varied range of forms, there is a common body plan to all members of this group. They are often referred to as the 'macropods' from the scientific name of the superfamily, which translates from the Greek to mean 'big-footed' (Egerton 1997). This term describes one of their most defining and distinctive characteristics; long toes on big feet, on the end of strong hindlegs. In contrast, the arms and hands are generally much shorter and weaker.

All of the extant species belong to two families; the smaller, more primitive potoroids (rat kangaroos, potoroos, bettongs), and the generally larger and more recently evolved macropodids (kangaroos and various wallabies). All living macropods are essentially herbivorous, the smaller ones being a little more omnivorous, and the larger ones being more standard grazers (Stahan 1998). The tropical rainforest and it's edge supports a diverse range of very different macropods, from the primitive rat kangaroos, to creek dwelling swamp wallabies, to rainforest floor pademelons, to tree kangaroos in the canopy.

Rat-kangaroos, Bettongs, and Potoroos
Family Potoroidae

Of the 2 families of kangaroos, the potoroids are considered the more primitive. Kangaroos are thought to have evolved from a possum-like ancestor. Thus the potoroids retain more 'primitive' arboreal possum-like conditions than their larger wallaby and kangaroo relatives, such as less developed hind legs and shorter feet than their 'big-foot' cousins, and slightly prehensile (climbing) tails. Potoroids are much more omnivorous in their diet, and they cannot generally digest grass like their larger stomach relatives, but eat 'easier' food such as fruit, tubers, fungi, insects and may even scavenge dead animals (Strahan 1998). There are also behavioural differences; unlike the bigger kangaroos, which essentially plonk down on the ground or rest up against a tree to snooze, the smaller, more vulnerable potoroids make nests (Strahan 1997), constructed out of plant materials such as grasses and fallen branches. They use their grasping 'possum-like' tail to carry material for their nest making; the glimpse of a small kangaroo bounding through the undergrowth with a curled tail full of sticks is quite an odd sight. Apart from reproduction and raising young, potoroids are not considered gregarious, and are usually seen scampering alone along the ground. Currently, there are about 10 species of potoroids in Australia (Strahan 1998).

Musky Rat Kangaroo *Hypsiprymnodon moschatus*. Smallest of all kangaroos. Rabbit-like in size and movement. Coloured chocolate brown with rusty shoulders. Restricted to floor of tropical rainforests of north-eastern Australia. One of Australia's two strictly daytime active marsupials. Unique within the superfamily for their bounding rather than hopping gait, and their habit of usually having two young at a time (rather than one). Tail can curl around to carry nesting material. Best places to see are Heritage Lodge, Cassowary House, Cathedral Fig, Red Cedar and Lake Barrine upper track

Family Macropodidae

Tree-kangaroos
Dendrolagus spp.

Kangaroos may have evolved from possum-like marsupials that once lived in trees; but the tree kangaroos represent a group of kangaroos that have *gone back* and re-colonized the canopy. While some species are noted on the ground in New Guinea, they are true tree-dwelling (arboreal) kangaroos and thus the largest animals to be found in the rainforest canopy. They are the only kangaroos that regularly move the legs independently, obviously useful in walking along branches and climbing up trunks. They also have large and sharp claws, and much heavier, thicker forearms than other kangaroos. They browse on a variety of different leaves of rainforest trees. However, they are not perfectly adapted to their habitat; their long tails hang down and are not prehensile (although they are useful for balance), and they often climb down backwards and rather awkwardly out of trees, or if alarmed, they may simply jump out and hop away, creating a crashing sound as they do so. There are 8 species of Tree-kangaroo, including some that have only recently been discovered. They occur mainly in New Guinea, but there are 2 species found in Australia, where they are restricted to the rainforests of north-eastern Queensland.

Lumholtz's Tree-kangaroo, *Dendrolgaus lumholtzi*, (pictured below) found mainly in rainforest of the Atherton Tablelands, usually more common at higher altitudes, and **Bennett's Tree-kangaroo *Dendrolgaus bennettianus,*** found from the Daintree River north to just south of Cooktown, where it is probably most common in gallery forest.

Wallabia spp.

This genus contains only 1 species.

Swamp Wallaby *Wallabia bicolor*. Medium sized wallaby with distinctive dumpy, pear-shaped figure. Short, thick dark hair, and thus sometimes known as 'Black Wallaby'. Distinctive face, with black muzzle and white stripe along cheek. Unlike many other macropods of similar size and larger, is usually quite solitary and mainly spotted alone. Nocturnal, but sometimes seen during day. Browses on variety of plant foods, shrubs to grasses, including ferns (Menkhorst and Knight 2001). Associated with range of habitats in it's distribution along entire east coast of Australia, including rainforest.

Pademelons *Thylogale* spp.

The unusual common name of 'Pademelon' is probably corrupted from a word used for this wallaby by Aboriginals in the Sydney region (Egerton 1997). They can be considered a rainforest wallaby, for the 4 species are found in the denser, wetter forests of eastern Australia and New Guinea, as well as in nearby sclerophyll forest and grasslands (Walker and Nowak 1985). Pademelons are rather small, squat wallabies with soft, thick fur and relatively short tails. They usually sleep and rest deeper in the forest through the middle part of the day. In the afternoon, night and mornings they may come out onto the edge of the forest and into clearings to feed, often wearing out little trackways in the denser foliage (Lindsey 1998). They feed on fruits, grasses and leaves and some have even been blamed for crop losses (Egerton 1997). Pademelons are usually thought as solitary animals, with feeding in groups only due to sharing of the resource, and are reported to scatter when alarmed rather than bunching together as the more gregarious macropods do (Lindsey 1998). However, this is countered by observations by John Chambers of the Red-legged Pademelon *Thylogale stigmatica* in North Queensland that seem to operate in groups, complete with dominant individuals and guards that sound perimeter alarms by thumping the ground with their big feet much like rabbits. Breeding occurs all year, with males that are not usually much taller than the females, but may weigh almost twice as much due to more muscle (Egerton 1997).

Red-legged Pademelon *Thylogale stigmatica*. Upperparts dark brown in contrast to paler underbelly, dark nose and paws. Rusty tinge to face. Also rusty tinge to limbs, which common name is referring to. Usually a pale to yellow stripe on outer thigh. Found in the wetter forests of eastern Australia down to NSW, and in New Guinea. Range has been reduced somewhat due to clearing of rainforest habitat for agriculture (Walker and Nowak 1985). But still common and can be observed every night in their loose groups at Chambers Rainforest Apartments.

Subclass Eutheria

Order Rodentia

It would be hard to argue against the idea that the rodents are the most successful of all mammals. There are about 2000 species worldwide, with at least 60 species in Australia. This represents, like the bats, almost a quarter of the mammal species of the continent. Rodents have also been successful in colonising practically every corner of the planet, and the tropical rainforest of Australia is no exception.

Rats and Mice
Family Muridae

There appears to have been at least two main invasions of rats and mice into Australia, and this is often represented by two different sub-families. There is a group of rats that reached Australia millions of years ago, and are thus known as the 'old endemics', as opposed to members of the other subfamily Murinae, that invaded the continent much more recently (Strahan 1998). They have thus had time to evolve into a distinctive group of species, including the sometimes quite large 'tree rats', the rare 'stick-nest' rats, the 'pebble-mound' mice, the cute 'hopping-mice', the 'water rat', and the melomys.

Water-rat, *Hydromys chrysogaster*. Often mistaken for platypus, as shares similar habitat and habits; is only other aquatic mammal in Australia. Genus name means 'water-mouse'. Australia's largest rodent (Cronin 2000). Long straight tail, pale for last third. Colour of body varies from blackish to brown to grey, with paler underparts, and sometimes fur has golden tinge; specific epithet means 'gold belly' (Strahan 1995). Fur is dense and water-repellent (Olsen 1998), and always seems sleek and wet from its swimming habit. Has large, partially webbed back feet. Can often be seen hunting during daylight hours, and in few times it is seen out of water, moves with bounding loping motion, appearing rather slinky and weasel-like. In water and on edge, feeds on various prey, including invertebrates, fish, frogs and waterbirds (Cronin 2000). Food is often taken to 'feeding platforms', and resulting scat is often full of scales, bones and has fishy smell (Triggs 1996). Hunted by birds of prey, snakes, cats, and by humans for pelt (Olsen 1998). Inhabits creeks, wetlands and estuaries throughout eastern and tropical Australia and New Guinea (Olsen 1998).

Giant Naked-tail Rats
Uromys spp.

Uromys are a group of 7 species of rodent found in tropical Australia, New Guinea, and other surrounding Pacific Islands (McDonald 2006). Most species are large, with the tail being even longer than the body, and is almost hairless and covered in scales. They have fairly generalized dentition and most species are reasonably omnivorous. There are two species found in Australia, and both inhabit the tropical rainforest. One species, the **Masked White-tailed Rat *Uromys hadrourus***, is relatively rare and confined to just a few higher altitude spots, while the following species is quite more common.

Giant White-tailed Rat *Uromys caudimaculatus.* As name suggests; a giant rat, with last third of tail being white. However, much more attractive than it sounds, with cute pink face and happy little jaunt. Commonly seen at night around tropical rainforest, often seen running across road on summer nights. Also quite arboreal, climbing vines and trees. Very efficient at gnawing into things, such as backpacks, to get at food,. Can even chew through tins, and not uncommon to find coconuts on lowland beaches with large holes that have been chewed out by these giant rats.

Melomys spp.
Mosiac-tailed Rats

There are over 20 species of these rat-sized mice, and they are found in eastern and tropical Australia, New Guinea and associated islands (McDonald 2006). Their tails appear naked and covered in scales. They are also quite arboreal and may be seen scampering around the tops of lodges in the rainforest at night. There are 4 species in Australia, and 3 of those can be found in tropical rainforest.

Grassland Melomys *Melomys burtoni*. Mainly found in woodlands, farmlands and mangroves of eastern and northern Australia, but also seen on tracks and roads on edge of rainforest. **Fawn-footed Melomys *M. cervinipes*.** Usually found deeper in various forests on coast, north NSW to north-east Queensland. **Cape York Melomys *M. capensis*.** Found in forest and woodland, northern Cape York.

Bats
Order Chiroptera

Bats are one of the most successful groups of mammals in the world, and in Australia there are at least 70 species; this is almost a quarter of this continent's mammals. Bats are nocturnal, flying and feeding mainly at night, and hanging upside down when they are at rest by day.

They are of course the only mammals that can fly and are best distinguished by their two leathery wings that stretch between the limbs. The fruit bats lack tails, but the smaller species usually have some sort of tail, and the type of tail it has assists in determining the family of bats to which it belongs, for example, there are the 'sheathtail bats', the 'freetail bats' etc. The wings stretch from the arms to the sides of the body, and there is a membrane stretching between the legs and tail. There is only 1 claw on each forelimb (Robson 2002). Many people around the world fear these bats for various reasons, and many of these reasons are unfounded, but one has to admit that when examined in the hand many smaller bats do often have rather grotesque little faces with various folds in the skin, wrinkly lips, squinty eyes, huge ears and facial hair.

Until very recently bats were placed into two different suborders; the Megachiroptera and the Microchiroptera. Late last century there was an interesting debate with some evidence and conjecture suggesting that Megachiropteran bats may be more closely related to lower primates than to the other bats (e.g: Pettigrew 1988). However, The most recent genetic evidence suggest these divisions are no longer relevant, and there is in fact some overlap between the two groups, with the 'megabats' related to the horseshoe and leaf nose bats (Jones 2006).

As to be expected from such a large order, bats vary enormously in size, form and behaviour. The largest, an tropical Asian species, has a wingspan of over one and a half metres (Strahan 1998). Most of the Insect Bats (smaller picture right) are usually quite small, and in fact include some of the smallest mammals in the world. The smaller, faster microbats are mainly insectivorous, and use echolocation to chase and catch their prey. The larger megabats usually eat fruit and nectar, and thus most species lack echolocation. The megabats tend to have larger bulging eyes and a cute fox-like appearance, while the smaller micro bats usually have tiny eyes and the various folds and flaps on their face may make them rather less attractive.

While most bats are insectivorous, this huge and diverse group also includes species that have evolved to feed on fish, fruit, nectar, blood and even other bats. For the majority of species, a special sense is used to catch their small and fast moving prey.

Echolocation

Bats produce their high frequency calls from the throat, and project them out from either the mouth or the nose. They then detect the 'ultrasounds' as they reflect off other objects. They use this echolocation sense both to navigate in the dark and to locate their prey. The 'echoes' can provide information on the size and distance of objects, and even their texture and speed. And they can catch prey as small as mosquitoes (Robson 2002). Bats can emit over a huge range of kilohertz (kHz), with an average of about 50 kHz. This is way above human hearing, as we hear between 3 and 18 kHz (Churchill 2002), so we tend to catch just one end of their calls, with younger, smaller humans tending to hear more calls. Different bats use different frequencies, depending on the species of bat and it's habitat. The higher frequency of the wave, the shorter it's length, therefore the greater the resolution. Some bats are loud 'screaming' bats over large distances in open areas. These tend to be lower frequency calls and therefore longer wavelength, so they have less resolution but can travel further. Others are 'whispering' bats, who emit high frequency calls over short distances with a higher resolution and can catch smaller prey and fly in closed environments, such as tropical rainforests.

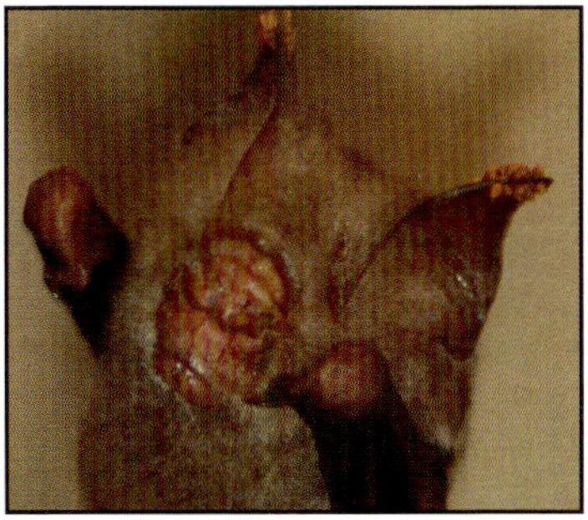

The strange folds and noses of bats are probably an aid in both emitting and collecting the 'echoes' of their calls. There can be 'horseshoe', 'lower leaf' or 'upper leaf' variations on the nose. The three families of bats with noseleafs emit the calls through their nostrils, while the more simple faced bats emit the sounds through their mouths (Churchill 1998). The 'tragus' is a piece of cartilage growing from the base of the ear and it may aid in giving directionality in detecting echolocation; their size and shape also assists in identification of species.

Like many animals such as frogs and birds, their calls are in fact species specific. So biologists can use computer analysis of calls to try and recognize species. The calls are detected and collected with machines, such as the 'anabat'. Calls can then be expressed as a sonogram, a visual representation of the call on the computer screen.

Bats in the Australian tropical rainforest

As with many groups of animals, there are many more species of bats in the tropics, and the forests of tropical Australia are no exception. Thus the highest diversity of bats is found in the north-east of the continent, with both fruit bats and insect-eaters being well represented in the Australian tropical rainforests.

Suborder Yinpterochiroptera

This recently erected suborder contains the group previously known as the 'mega bats' as well as the 'false vampires' and the leafnose/horseshoe families.

Flying Fox/Fruit Bats
Family Pteropodidae

There is a total of over 40 genera and over 160 species in this family. They are largely restricted to the more tropical forests of the world, but of these warmer regions, they are only found in the 'Old World' of Africa, Asia, Australasia and the South Pacific. They are notably missing from the Americas; there, some of the smaller microbats in the tropics have filled in similar niches and are even locally known as 'fruit bats' (e.g; Emmons and Feer 1990).

Those species that live in Australia have likely evolved from Asian stock. Their powers of flight have meant they are one of the few placental mammal groups that could invade the isolated Australian continent millions of years ago.

The bats in this family generally don't have echolocation; only a few African and Asian species can echolocate, and this is actually performed in a different way from the micro bats, with the clicking of the tongue (Robson 2002). The lack of echolocation means megabats do not have the facial folds of the smaller insect eating bats, and thus have much simpler and more attractive fox-like faces. They have well developed senses of sight and smell, and these are used in locating fragrant flowers and ripening fruit. Unlike the other bats, they tend not to have tails, and have two claws (rather than one) on their forelimbs (Robson 2002).

Fruit bats are now recognised as vital pollinators and seed dispersers of many rainforest plants. The so called 'fruit bats' are not really fruit *eaters*, but, rather, fruit *suckers*. The digestion of the actual fibres of fruit would require a larger stomach, a physical feature that such flying animals would rather do without. They chew on the fruit and suck out the juice, spitting out the fibrous parts. (This liquid pulp goes through their system relatively quick, so the quick digestion and spitting are both good reasons not to stand underneath feeding flying fox!)

They are generally attracted to the more odorous fruits, as this is how they are located during the night when they feed. Flying fox find many of the same fruits that humans also find attractive; some of the fruits we now grow in the tropics, such as mangoes and bananas, had their wild ancestors dispersed in the wild by these bats. Unfortunately for human/bat relationships, this means orchards of these fruits become the target of these bats.

Pteropus spp.

At first, these large bats alarm many people, due to their large size and harsh cackles and cries. However, a closer inspection reveals a rather more attractive view; the bats are given the name of 'flying fox' due to their intelligent, fox-like face.

Like the other bats, flying fox feed by night and rest by day. During the day they roost in large groups. Some of these 'camps' can be in quite open areas; their very presence often ends up defoliating the trees they hang in, opening up the area even more. The bigger populations tend to be noisy and smelly places; they seem to spend much of their "night" squabbling and interacting, and flapping their wings. This latter behaviour assists in keeping cool in the open heat of the day in the tropics. Some of the biggest camps have developed in major urban centres, such as Cairns in North Queensland and Madang in northern New Guinea. In fact, the colony of Spectacled Flying Fox in the middle of urban Cairns is the largest of this species anywhere (Whybird and Clague 2001).

Spectacled Flying Fox *Pteropus conspicillatus* Tropical rainforest equivalent and replacement of Black Flying Fox of northern Australian savanna and mangroves; looks very similar from distance. All black, except for yellowish shoulders and neck, and yellow colour fur around eyes forming the "spectacles' of common name. Found in and around tropical rainforest and closely associated habitats such as mangroves, paperbarks and gardens in north-east Australia; also in New Guinea. Easy to observe within this restricted range; occasional camps along Daintree River and on Mardja boardwalk, and large more or less permanent camps at Tolga, and in middle of urban Cairns, where they can be seen leaving their camp and flying off every evening.

Blossom Bats
Subfamily Macroglossinae

These bats are relatively small megachiropterans, (pictured below with a human thumb for scale), being only slightly larger than many microbats. They are light brown, sometimes almost rufous. In flight, they look like a larger insect bat, but when hanging they look just like a miniature flying fox, with their long, fox-like snouts and large eyes. As the common name suggests, Blossom Bats have evolved to feed mainly on nectar from flowers. The scientific name *macroglossus* means 'large tongue', and as well as being long, it is flexible and has a brushlike tip. They also ingest insects during this feeding, and after visiting the flower they will lick the pollen that gets in their fur, as well as eating fruits and even leaves (Law and Spencer 1998). Blossom bats are found in a range of habitats where there are large nectar producing flowers; including rainforest, mangroves, woodlands, heath, orchards and gardens. They range from South-east Asia, through New Guinea and various Pacific islands, to Australia (Flanney 1995a). There are 2 species of Blossom Bat found in Australia.

Common or **Eastern Blossom Bat *Syconycteris australis*** and **Northern** or **Least Blossom Bat *Macroglossus minumus*** (pictured below). Both very similar looking; former more rufous in colour, while latter has longer snout, tail stub and skin flap along inside of legs.

Leaf-nosed Bats
Family Hipposideridae

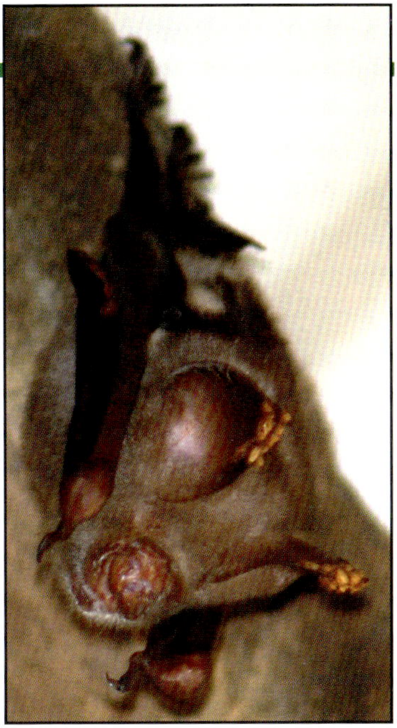

There are over 80 species in this family (Jones 2006), with 7 species in 2 genera found in Australia (Churchill 1998). One genus, the endemic *Rhinonicteris*, is not represented in the Australian tropical rainforest, but the other genus *Hipposideros* does have several species in this habitat, including the **Diadem Leaf-nose Bat *Hipposideros diadema*** (pictured right). The species in this family are usually known as the "Leaf-nosed Bats' due to the elaborate shape of their nose flaps, however, the family and main genus name actually means 'horseshoe', which makes for some confusion, as the bats commonly known as the "Horseshoe Bats" (due to the even more elaborate shape of their nose leaves), are in a different family, the Rhinolophidae.

The face of the Leaf-nosed Bats is distinctive; from a distance it looks like a little 'piggy' nose. But on a closer look there are two nose leaves; the lower one is larger, rounded in shape, and fairly flat with 2 large nostril holes, while the upper leaf is smaller and usually has a series of folds or indentations in it. The ears are usually very large, and often have fine horizontal lines, with no tragus. Their tail are inconspicuous, with the tail membrane completely enclosing the tail and usually folding up over the bats rump when hanging (Churchill 1998).

Horseshoe Bats
Family Rhinolophidae

The small family includes only the 'horseshoe bats'. They are similar to the Leaf-nose bats in the previous family in their large ears, inconspicuous tail and elaborate nose structures. However, their facial folds go even further, with the upper nose leaf extending into a high triangular tip, and the lower leaf in the shape of a 'horsehoe'. There are only 2 to 3 species in Australia; two have distributions that include North Queensland and can be seen in caves in rainforest and monsoon forest.

Eastern Horseshoe Bat *Rhinolophus megaphyllus* (pictured left) has a more rounded 'horseshoe' leaf, while the rarer **Greater Horseshoe Bat *Rhinolophus philippinensis*** has a more angular horseshoe lower leaf and even longer ears and forearms.

Suborder Yangchiroptera

Sheathtail Bats
Family Emballonuridae

There are over 50 species in over a dozen genera worldwide in this family (Jones 2006), with some 8 species found in Australia, and most being found in the tropics (Menkhorst and Knight 2001). The most distinctive feature of the sheathtail bats, and the feature that gives them their common name, is the way the tail projects out through the upper surface of the tail membrane about halfway along.

Free-tailed Bats, Mastiff Bats
Family Molossidae

This family has 16 genera and almost 100 species worldwide (Jones 2006), with 11 species found in Australia (Churchill 1998). The name of the family is derived from the tail sticking and uncovered by the tail sheath for half of it's length. They only have a relatively simple face compared to many of the other bats, with a broad cut-off muzzle and wrinkled jowls/lips, resulting in the appearance rather like an little ugly dog (pictured left). The ears are sometimes connected to each other by skin at the base, and there is a small tragus (Churchill 1998). The legs are short, but thick and strong, with long hairs sprouting from the claws. The species in this family are found in a range of habitats, but are mainly tropical and subtropical (Wilson 1997).

Family Vespertilionidae

With over 400 species, this is the biggest family of bats in the world, and they are found almost everywhere (Jones 2006). There are at least 35 species found in Australia (Menkhorst and Knight 2001). The bats in this family tend to have much simpler faces and smaller ears, although there are exceptions. Their tails are enclosed in the tail membrane.

Large footed Myotis *Myotis adversus* (pictured below). Very large clawed feet. They use these strong feet to catch small animals from the surface of waterways. Found around water, eastern and northern Australia.

Bent-wing Bats
Miniopterus spp.

These small bats are usually coloured a dark chocolate brown. They have a distinctive high domed forehead, and small broad ears. These bats are one of the more commonly seen bats around human inhabited areas, often seen flitting around street lights and roosting in buildings in and around rainforest areas.

Even-toed Ungulates Order Artiodactyla

This famous order contains many of the most spectacular mammals on the planet, including pigs, hippos, camels, llamas, deer, antelope, goats, sheep, bison and cattle. They are the among the largest and most dominant mammal herbivores; on every continent except two; Antarctica, and Australia. While cattle occasionally wander into the Australian rainforest, there is one species from this order that has become a major force in this habitat.

Wild Boar *Sus scrofa*. Large, dark. Domesticated into the pig thousands of years ago; farm animals that escaped have become wild animals again. Evidence of their destruction easy to see, especially at lower and mid altitudes, with massive soil diggings, uprooted seedlings eg; Marrdja boardwalk. May be seen running across the Cape Tribulation road. One of few potentially dangerous rainforest animals; intimidating to come face to face in the forest. Should be avoided.

Order Carnivora

This order contains the many familiar placental carnivores, including bears, racoons, tigers, lions, hyenas and wolves. Placental carnivores are the dominant terrestrial predators on every continent, except one - Australia. The order includes several animals that have been introduced into Australia and caused havoc, including cats and foxes. The very first introduced animal was the Grey Wolf. Brought into Australia from Asia by humans several thousand years ago, the domesticated version of the Grey Wolf, the Dingo, likely lead to the extinction of the two biggest predators Australia had at the time - the Thylacine and the Tasmanian Devil. Sadly, because of the time passed, we may never know how many other smaller to medium sized native animals were effected by this animal. The Grey Wolf, Dingo, Feral Dog, and domestic dog are all the same species. Today they continue to kill native animals, with feral and domestic dogs killing the largest of all native rainforest animals, the endangered Cassowary.

Humans
and the rainforest

Early humans of the rainforest

Arrival

Evidence from direct scientific dating techniques suggests that humans have been in Australia for at least 40,000 years. Further, indirect evidence from fire change to the landscape is used to suggest even longer periods of occupancy. Over this long time, there may have been several waves of successive groups of humans coming into Australia. Evidence of occupancy from Mt. Mulligan, on the western edge of North-east Queensland, suggests people have been in the local area for at least 37,500 years (Bottoms 2000). Dates from within rainforest areas are harder to come by, with physical evidence such as rock paintings being far less obvious and common than in the adjacent savanna. Changes in vegetation based on pollen samples in Lynchs Crater on the Atherton Tablelands have been put forward as evidence of people changing their habitat some 45,000 years ago (Bottoms 2000).

Culture and language

Today we often tend to lump all native Australians together as 'aboriginals', for convenience in political and economic situations, and in historical and cultural generalisations. However, this all-encompassing name tends to ignore the diversity of cultures within the huge area that is the Australian continent. It is estimated there were over 200 languages and more than 700 dialects prior to white settlement, reflecting a cultural diversity comparable today to an area such as Europe. Thus, the areas of north-eastern Queensland and Cape York were home to many different language groups. The following language groups were roughly thought to inhabit the following areas:

Kuku-Yulangi people of much of the Daintree, from Mossman to Bloomfield,
Tjabukai the people around and north of Cairns,
Yidiny and Dyirbal , south of Cairns and much of the Atherton Tablelands
Wargamay, between Cardwell and Ingham (Bottoms 2000).

Below - A member of the Kuku Yulangi at Mossman Gorge showing some rainforest fruits.

The ecology of Aboriginal people

The aboriginals of the rainforest not only had language differences, but were also recorded by early anthropologists to have distinct physical differences to their more open country neighbours. They were observed to be smaller and darker in stature, and these are traits that are often found in forest peoples around the world, and indeed are found in many different species of mammals of darker, closed habitats.

As with many Aboriginal groups around Australia, movement in response to different seasonal conditions and resources was probably fairly common. Many people were known to move up and down the coast, in and out of the forest into drier woodlands, and from the coastal lowlands to higher altitudes and then back again.

The tropical north-east of Australia was noted for it's comparatively high density of people (Bottoms 2000), likely due to the large range of adjacent habitats, including the edge of the rainforest, savanna woodlands and grasslands, paperbark wetlands, mangrove forests, mudflats and coral reefs. This range of habitats allowed for a variety of quality food species, from fish to flying fox (pictured below).

The deep interior of the tropical rainforest may not have been as densely inhabited as these adjacent habitats all year round. It is difficult for groups of relatively large animals, such as humans, to get enough food to support themselves in the rainforest, for large game such as cassowaries and pademelons are shy and of low density. Much of the protein was probably obtained from aquatic organisms, the creeks being fished for turtle and large fish such as jungle perch.

Bush tucker and knowledge

The pattern of organism distribution in the rainforest is of high diversity of species, but low abundance of individual animals and plants. Thus, to survive on a diet of many different species in low numbers meant the inhabitants had to accrue considerable and intricate knowledge of the rainforest, its species and ecology. This was reflected in the widely recorded use of seasonal timings and 'indicator' species; that is certain species would indicate when another resource was available. For example, the flowering of the October Glory vine indicated in some areas that the Brush Turkey eggs could be dug up from their mounds.

Many different types of fruits in the rainforest are today known to be edible, with this knowledge often coming from early explorers and their aboriginal guides. Some are easily and obviously edible, such as many of the *Syzygium*, but other plant foods can not be eaten by people today and are considered toxic (Roberts *et al* 1995). However, the original people of the tropical rainforest were particularly noted for the amount of toxic fruits in their diets. They managed to eat these toxic foods only through elaborate processing. They were sliced into parts or mashed up, and then stored in permeable bags in running water for days until the toxins were leached out. Such plant foods included black beans, cycads and many in the laurel family; some 14 species were recorded all up, and some of these were wet season staples (Bottoms 2000).

European exploration

It was likely there was contact between the people native to the north-east coast and people living to the north in the islands of the Pacific and in south-east Asia. The first 'official' recording of the tip of the Australian continent, Cape York, was by a Japanese pirate Yamada Nagamasa around 1600, who made his living stealing from Europeans trading in the area. He eventually found enough pearls to retire a rich man back in Japan. Just a few years later, the Dutch would arrive.

This would be left to James Cook in the *Endeavour,* sent from England on his first major voyage. In 1770, while negotiating the 'labyrinth' of coral that would later become known as the Great Barrier Reef, the *Endeavour* was severely damaged. Cook named the headland he had seen on the previous day 'Cape Tribulation' as this was when all their trials and tribulations started.

Competition between the rising powers of Europe resulted in explorers not just from Holland, but also Spain, France and England sailing the general region. The Dutch explored and colonized the tropical rainforest of the spice islands to the north, throughout what is now known as Indonesia. However, in their exploration of Australia (which subsequently became known as "New Holland" for several decades), they generally encountered the drier areas of the western and northern coasts, including the west coast of Cape York in 1606. They were not aware of the strait and presumed it was connected to New Guinea to the north (they would have been right if they had been there just 12,000 years earlier). The Dutch were famously unimpressed both by the sparse vegetation and the 'miserable' natives of the north and the west, and they didn't make it to the wetter forests of the north east.

Cook and his crew needed to find a safe place to repair the ship, and as they searched, they passed the rainforest of Cedar Bay and the Bloomfield coast, until they finally ended up on the banks of what is today called the Endeavour River, in a place that would later be called 'Cook's Town'. While this enforced landfall was a major hurdle to the expedition, it also resulted in many of the first recordings of Australian animals and plants for western science. The well connected and rich 'gentlemen botanist' Joseph Banks had paid a lot of money to organize a natural history party on board the Endeavour. Included was the scientist Solander, and several artists such as Parkinson. They recorded over 200 plant species new to European science, many of them rainforest plants. Among the rainforest animals, they saw their first flying fox, as well as more open country animals, such as dingo and kangaroo.

For the next century as the penal centres in south-eastern Australia expanded, the tropics remained a remote place. Exploration was mainly by sea. One of the most interesting visits was one of the earliest: the infamous Bligh pre-dated many of the later famous explorers, but not by choice! In 1789 the mutiny on the *Bounty* resulted in him being placed into a long boat and put out to sea. With 18 men, the small boat travelled thousands of kilometres from Polynesia, through the Great Barrier Reef, and made first landfall at Restoration Island, near the modern day rainforest of Iron Range National Park, Cape York.

Later explorers who sailed past the rainforest of the north-east included Mathew Flinders on the *Investigator* in 1801 and Jeffereys on the *Kangaroo*. Philip Parker King followed in several expeditions; in 1819 he sailed the *Mermaid* into the mouth of the Escape River near Cape York, and was nearly destroyed on a sandbank at the mouth, his narrow escape giving the name to the river. In 1848 Owen Stanley captained the *Rattlesnake* with naturalist John Macgillivray, accompanied by the *Bramble*.

The sheer thickness of the rainforest, known as the "scrub', prevented the early explorers from penetrating far into the land. This was tragically confirmed by the infamous expedition led by Kennedy. In 1848 he started off in the Mission Beach area with 12 men, 27 horses and 100 sheep. Almost immediately they ran into problems negotiating the thick vegetation. His aim was to reach the top of Cape York, but Kennedy himself died, and no animal or person made it to the meeting point at the end, except for one, an aboriginal guide commonly known as Jackey-Jackey. After the failure of Kennedy, there was continued reluctance to penetrate into the thick scrub of the coastal and mountain areas. In 1864 the Jardine brothers made it to Cape York overland with 10 men and 250 head of cattle, but they went inland up through the more open woodlands and grasslands rather then the thick scrub. Thus for many of the early explorers and settlers the scrub was to be avoided, and it needed to be fired, chopped down and cleared as much as possible to 'open up the land'. There were only a few that explored the forest on it's own terms.

There was only a handful of explorers that interacted with the native people and penetrated the rainforest. One of the more controversial of these explorers was Christie Palmerston, who was the first white man to climb Bartle Frere. He wisely used the local aboriginals as guides, but was certainly not shy of shooting them. He would open up many trails between the mountains and the lowlands, and is honoured today with the naming of the Palmerston highway, a major road from the coast to the tablelands.

Among those exploring the rainforests of the region were naturalists and scientists. Chief among those that was zoologist Carl Lumholtz, born in Norway in 1851. He came to Australia in 1880 to explore the tropical rainforest of North Queensland. He several years living with the aboriginal people in the forest and from this he would gain insight in many of the last of the wild people of the forest, and also see some rare animals, some new to science. He started roughing it in the forest, taking the basics of salted meat, wheat flour and sugar and only ate what he could shoot and enlisted the help of aboriginal people by trading their services for tobacco (Pizzey 1985). He made observations of rainforest birds, including the fact that the male cassowary looked after the chicks. More significantly he found several species new to western science, including a greenish possum that later become known as the Green Ringtail. His Aboriginal guides also caught an animal known as a *boongary*, a tree climbing kangaroo known that would later be known as the Lumholtz Tree Kangaroo. Lumholtz published a book of his experiences called *Among Cannibals* in 1889.

Edmund James Banfield, was one of the first modern-style naturalists to actually live in the region. He was born in England in 1852, but came to Australia as a child. He worked as a journalist for a paper in Townsville, and ended up virtually running the paper, until 1897 when his doctor gave him less than 6 months to live unless he changed his lifestyle. So he and his wife "dropped out" and settled on the then fairly remote Dunk Island; he would ultimately live in good health for another quarter of a century until 1923!

Above - the grave of Edmund Banfield is on Dunk Island today.

During those years he wrote several books for the visiting tourist to North Queensland, and he popularized the idea of the romantic idyllic existence on rainforest tropical islands through a series of columns for the two local papers. Eventually his writings became published in England and would become popular throughout Australia, but also other parts of the world, including the first and most famous *Confessions of a Beachcomber* (1908), which has almost continuously been in print until today, followed by *My Tropic Isle* (1911), *Tropic Days* (1918) and the posthumous *Last Leaves from Dunk Island* (1925). While living on Dunk, he made various observations on the birds and other animals of the tropical rainforest there, such as the Pied Imperial Pigeons and other birds (Pizzey 1985). He also made observations on the aboriginal people of the island; this was at an interesting transitional time when they still lived traditional lifestyles, but were starting to use more modern European introduced elements, such as blankets instead of those carefully crafted from the bark of trees, or a piece of bent wire instead of a slowly and skilfully fashioned shellfish hook (Banfield and Noonan 1994).

Early European settlement and exploitation

The potential of the settlement of North-east Queensland was initially triggered by reports from George Dalrymple, who had been sent to explore the area by the Queensland government in the early 1870's (Ricthie *et al* 1995). The previously mentioned Palmerston would also be involved in the discovery of gold in some of the smaller remote rainforest creeks. However, everything changed much more dramatically when gold was discovered in large amounts in the river systems of the dry country to the west of the rainforest. This brought thousands of people into the area in a series of short but intense gold rushes during the 1870's. This directly led to the setting up the settlements that form the basis of the main towns in the region today, including Port Douglas and Cairns.

Timber

When Europeans first began to explore the area the immense jungle understandably suggested an infinite supply of timber. However, the huge diversity of species made it difficult to determining which timbers were which. Many different species were discovered and classified by the foresters, and received what are today considered strange and sometime inappropriate names after the properties of the wood that we cannot today see in the wild. Thus trees were called 'Silky Oak' for species that are not all related to oaks, (and don't feel particularly silky!). Ultimately, a few dozen timbers were focused on, with timber getters cutting through the forest to get at recognized timbers. These trees included various of the 'Silky Oaks', the best known probably being *Cardwellia sublimis*, the 'Northern Silky Oak'. However, the most valuable tree was the 'Red Cedar', a tree already known from other rainforests south of the region. It provided a water resistant timber that was in such popular demand for boat building and furniture, it was often nicknamed 'Red Gold'; it was at one stage reckoned to be, weight for weight, worth more than gold. Many of the forests that are today accessible to roads were thought to be selectively logged for this and other timbers. Today, many of the state forests and national parks are criss-crossed by old 'snig' tracks where the timber was hauled out; many of these are still used as access roads, some are used for walking trails, while others have all but grown over. Until just a few years ago there was a spectacular example of a Red Cedar at Gadgarra just near Lake Eacham, however, it has fallen over. There are still great examples of standing Kauri Pines at Lake Barrine, as well as examples of the various untreated and untreated rainforest timbers at the teahouse. There are also samples of timber and wood products at Tolga outside Atherton and in the Daintree Village.

Agriculture

After the various rushes for gold, tin and timber, a small number of people settled the area, continuing the clearing of land for agriculture. While the lowlands were mainly used for sugar cane and banana, the relatively rich soils and high rainfall of the tablelands were used for a huge range of products, including strawberries, corn, mangoes, avocado, coffee, tobacco, citrus and eventually a plant of the rainforest itself, the Macadamia (pictured right).

Sugar cane

Much of the coastal lowlands of north-eastern Australia have been cleared for sugar cane. It remains the single largest crop of the region, and by itself constitutes the largest industry after tourism. The plant itself is essentially a giant grass that can grow up to 4 metres tall before sending up collections of mauve flowers on tall stalks above the plant (pictured left).

This is around the time when harvesting begins, as it is when the sucrose has reached its peak in the plant. Until recently harvesting was done after burning the crop, but today it is mostly green-cane harvesting, using converted combine harvesters (pictured left). The machine chops the green tops off and then cuts the brown stalks into small sections (billets).

The cane must then be transferred to the mill within 16 hours, so the entire process is tightly co-ordinated. The sugar cane is loaded into large yellow containers, which then travel on railway tracks (pictured left).

The tracks lead to the nearest mill (pictured left). Here, the juice is squeezed out of the crushed cane. The actual raw sugar is then stored in bulk sugar terminals in towns such as Cairns. The use of sugar in other products requires further processing to produce the finer white powder sugar, and this is done at other plants further south.

Bananas

In North Queensland, bananas are second biggest crop after sugar cane. The banana we eat today as a fruit has come from either *Musa acuminata*, or a hybrid of that and *Musa balbisiana*. Both of these species are from the forests of Southeast Asia (Tidemann and Nelson 1987). The bananas that are grown on farms for human consumption have been radically changed from their wild ancestors over many years. While bananas in the wild have quite large angular seeds in the pulp, the seeds in cultivated bananas have been reduced to mere dots, or have not eventuated at all. In banana cultivars the fruits ripen without pollination. The new plants are grown vegetatively from suckers (Tidemann and Nelson 1987).

When visitors go past the banana plantations, they will often see bags covering the fruits. These have several different purposes, including protecting the fruits from their natural dispersers such as bats (or birds); only minimal damage to one fruit may set off the whole bunch to ripen. In the tropics, these cultivated bananas ripen all year (Tidemann and Nelson 1987). However, farmers generally pick them before they ripen, while still green to protect them from mechanical damage. They may then apply ethylene which artificially triggers off the ripening process in such fast ripening (known as 'climacteric') fruit (Tidemann and Nelson 1987).

Tea

Tea is grown in several locations of previous tropical rainforest in northern Queensland, including at the lower end of the Palmerston highway and Daintree Tea on the Cape Tribulation road (pictured right). Tea is a small shrub and in Australia appears in hedge like rows; this is due to the relative expense of labour and the leaves thus being machine harvested.

Dairy and Beef Cattle

While general beef cattle can run free on relatively poor land, dairy cattle needs higher quality fodder, and much of the cooler and richer volcanic soil areas of the Atherton Tablelands were cleared for dairy cattle.

The explosion of development

All this agricultural development resulted in estimates of 40% of the tropical rainforest being cleared in the last century. Most of the clearing occurred in the flatter lowlands closer to major centres, and across the undulating tablelands. The clearing began with agriculture, and continues with increasing urbanisation. The backpacker and then the eco-tourism booms lead to a increase in visitation, with many of these temporary visitors later coming back as permanent residents. Migration from the big southern centres led to the populations of Cairns and Port Douglas doubling in the last two decades. With that population increase comes the need for infrastructure, and with that development continued the biggest threat to the environment, more destruction of habitat. Rainforest in many areas around urban centres continues to be cleared for new residential developments. And along with all this increasing human population and urbanisation is the parallel accumulation of societies by-products and wastes. Unfortunately, there is little reason to think the process will be halted, as local populations are projected to continue to increase.

As their rainforest habitat is cleared and urban and farmed areas invade the forest, wildlife may be forced into the human habitat. Road kill is one of the more obvious consequences of this clash between two worlds. When larger, rarer species, such as tree kangaroos or cassowaries are run over or killed by dogs, their death is a significant loss to the population. Fast growing urban areas such as Cairns have seen very recent local extinction; less then two decades ago, there were half a dozen cassowaries living around Mt. Whitfield, right in the middle of urban Cairns; they are now all gone.

Conservation

Environmental awareness as a political force began to develop in the USA and Europe in the 1960's, and there were increasing populations of conservation-minded people living on the land, the 'hippies' or 'ferals', in North Queensland in the 1970's. It was about this time that scientists began to realize the extent in which this tropical rainforest was uniquely Australian, and not just invasive from tropical Asia. In the 1980's a road was proposed to follow a track connecting Cape Tribulation and Bloomfield by the then conservative and pro-development Queensland State government. This was known as the 'Bloomfield track', and it became the site of one of Australia earliest and biggest modern environmental protests. The protests began on a small scale with locals, and then began to become organized through state and national conservation organizations, attracting people from all parts of Australia from all walks of life, many who had never had the chance to see the area. Dramatic demonstrations included people chaining themselves to trees and being buried up to their necks in the ground in front of bull-dozers. The recent ideas involved in the realization of the areas biological significance were used in justifying protection of the region, and continue to be used today. For all their work, however, the road went through. However, the protests did get the region recognized nationally and internationally, and generated huge amounts of publicity for the area. Their actions lead to eventual further theoretical protection by declaring much of the remaining tropical rainforests of north-eastern Queensland under World Heritage status in 1988. This covered almost 900,000 hectares (Doyle 2000), including some 80% of the North Queensland rainforest, including the Daintree National Park, as well as other reserves west of the area, such as those that include the CREB track. These moves had a huge effect on previously extractive industries such as timber-getting in native forests.

Research

The latest and perhaps one of the more interesting industries that has arrived in the tropical rainforest is science and research. Previous to the 1990's much of the tropical research in the region was performed by scientists trained and working in southern universities, or based with national scientific organizations like CSIRO, or National Parks. But the establishment of a campus of James Cook University at Cairns has seen an increase of research scientists in the area, working on everything from plants to pythons to possums. Now, there is a generation of scientists being trained locally in tropical biology.

Techniques for researching the rainforest canopy

It didn't take long for western visitors to the tropical rainforest to realize how rich this ecosystem was. Further, scientists began to recognize that much of the life was not at the bottom floor of the rainforest where they where, but high up in the canopy, above their heads. While the canopy comes down to ground level on the edge of the forest at clearings, rivers, and disturbances, much of the canopy was well out of reach. Therefore, humans needed to get up into the canopy. To accomplish canopy research meant developing techniques to get humans up there efficiently and safely. At first, traditional rock climbing techniques and equipment was used, but this has it limitations. Other techniques were used, including giant nets dropped onto the canopy from aircraft which then allowed people to bounce around the canopy. Aerial walkways were then developed, including several famous ones in Australia, Malaysia and the Americas. However, they were limited in where they could go, and were very expensive. Today many of these walkways are used more for tourist ventures, allowing visitors to get up into the canopy. The latest technique for canopy exploration is the canopy tower, such as the one at Cape Tribulation (pictured below). This involves an expensive initial investment, as the set up begins with a large hole dug into the ground into which a single tall tower is planted. From this extends a horizontal arm, and from this hangs a cable at the end of which is a single cage. The arm can swing around, and the cage can be lowered up and down. This enables researchers to go up and down and around and the crane can cover a huge area in 3 dimensions. The operators can map the area that the crane covers, and then once they have programmed the locations of particulars trees into the computer, then they can be taken to any specific point in this multi-dimensional environment.

Wildlife research

Research is also conducted on particular species of plants and animals. This is conducted for various reasons; at it's most obvious, scientists are just asking basic questions about the animal and it's life cycle. Further study may take more complicated lines of enquiry, such as how it interacts with it's environment and other animals; it's ecology. More applied research may delve into the conservation biology of the animal; what is it's future in a changing climate, or how does it interact with humans?

For example, there has been and continues to be research projects done on the endemic rainforest possums. Studies have included ecological research on their diets, while others have focused on their physiology. Such studies may include the catch and release of animals, including the placement of radio collars (pictured).

Perhaps the most infamous of conservation biology issues in the tropical rainforest is to do with the decline of frogs. In the last few decades in different parts of the world, certain frog species have experienced drastic declines in what had been thought of as relatively pristine rainforest streams. By incredible scientific fortune, surveys of amphibians were actually being undertaken before, during, and after the 'crash'. The surveys measured limbs of the caught frogs and found evidence of asymmetry. This is a technique often utilized in the analysis of insect ecology to detect population stress. Scientists then undertook an international investigation that worked through a process of elimination and many different causes were researched, until finally a fungus was identified as the culprit. A cooperative study between James Cook University and NASA suggested there was also a correlation between the time just before the population crashes, and certain climate extremes. The culprit fungus that was found and identified may have actually already been naturally present in the frog populations, with the climate stress bringing it up to more threatening levels.

Other research may be just using the animal as a tool to ask 'big picture' questions. A good example of this latter research is conducted with common butterflies such as the Eggfly. At James Cook University in Cairns they are not just the subject of questions about the feeding biology or colours of the animal, but also deal with larger evolutionary issues such as sexual selection and male competition. To conduct controlled experiments, artificial habitats inside giant flight cages have been constructed, and butterflies are captured and marked as individuals for behavioural observations (see photograph).

Tourism

Rainforest tourism has existed on a local scale since early last century, with locals visiting places such as Lamington in south-east Queensland, and Baron falls, Dinner Falls (Hypipamee Crater) and the Volcano Lakes in the north-east. However, international and interstate exposure of the Australian rainforest was fairly low until publicity from the conservation battles in the 1980's over the Daintree. Australian and overseas tourism boomed (and busted a few times) and by the late 1990's tourists were estimated to be paying over $440 million annually for their accommodation and tours (Driml 2000). Further, this amount doesn't take into account the peripheral services for the visitor, from souvenirs to restaurants, or the tourism employed locals and connected industries, from printing houses to mechanics. The corresponding increase in economics has also seen increased development of the community and facilities. Today it is even possible to get a good cup of coffee in North Queensland.

Of course, it is a now known that tourism unchecked tends to repeat an ironic patter of growing quickly and destroying the very product it grew around. Thus, there are many economic, community and environmental management issues, at the interface between wilderness, wildlife and people. These issues have many sides; one of the concerns in ecotourism is the feeding of wildlife. This happens in lodges and tours in the rainforest, and it enables people to glimpse difficult to spot wildlife; however on the other hand, animals can get sick, dependant, or too used to people.

Another issue is quality of interpretation. This varies from signs on tracks written by national parks, to tour guides. There are courses at institutes such as TAFE for tour guides, and university level studies in ecotourism. As tourists themselves become better travelled, they expect better interpretation delivered by suitably qualified professionals. Thus, an increasing number of the higher quality and specialized tourist products use biologists as guides.

Places to see
Tropical Rainforest

Hinchinbrook island

Hinchinbrook Island is about 34 kilometres long, with many dramatic and craggy peaks and ridges, including the tallest peaks of Mt. Bowen and the 'thumb' at 1121 metres and the distinctive landmark of Mt. Diamantina and its squarish granite outcrop atop at about 955 metres. There is evidence of indigenous people on the island in the form of fish traps in the tidal areas. With only a narrow channel between it and the coast, it was not recognized as an island for some time by various explorers, including James Cook on the Endeavour in 1770, the Cyclops, Kangaroo, Mermaid, and the Beagle. Finally, the Bramble and the Fly confirmed that it was an island in 1843, some 65 years after Cook. The island was subsequently named for one of its peaks, Mt, Hinchinbrook. The island is reported to be the biggest single island national park in the world. It is almost all protected national park, with only a very small low key resort. The Thornsborne trail follows much of the length of the island, but only 40 people are allowed on the track at one time. You can visit the island via boat from the mainland and several companies cruise through the channels (without landing on the island itself) on their multi-day trips.

Mission beach

Mission Beach is visited by many independent domestic visitors, but it is relatively free of the hordes of international day trippers that invade the Daintree. Here, there is an excellent range of walks, from easy boardwalks to longer day hikes along the Kennedy track. There is a range of accommodations and eateries, and it is still the best place for spotting cassowaries.

Wooroonooran (Bellenden Ker/Bartle Frere) National Park

Mt. Bartle Frere and the closely associated Mt. Bellenden Ker are contained within Wooroonooran National Park. This park covers some 79,500 hectares, making it slightly bigger than the Greater Daintree National Park. The park contains a range of rainforest types, from the most complex lowland communities, to the quiet cloud forest of the peaks. Together, the mountains Bartle Frere and Bellenden Ker comprise the biggest mountains in northern Australia, with Bartle Frere being the highest peak in Queensland at about 1620 metres. Like many of the peaks in the area, the two mountains began as underground magma almost 300 million years ago. As the molten rock rose, it began to cool slowly, turning into dense, coarse-grained granite. However, the underground granite did not see the surface until millions of years later when the softer overlaying and surrounding rock was eroded away, leaving the mountains we see today. Many large boulders of similar intrusive origin can be seen in the valleys and creeks below. The area is wet, rugged, and thus seen little development. It is in fact Australia's wettest National Park. While there is no rain gauge on Mt Bartle Frere itself, the nearby Mt. Bellenden Ker (pictured below) is the wettest recorded place on the continent, with an average of over 8 metres a year (Tracey 1987). The rainforests around Bartle Frere have been subject to some small scale exploitation. Gold was searched for in some of the creeks in the area. There are various 'snig tracks' that were used as roads to pull out logs by the timber-getters. Today, few tour companies conduct walks in the area, and there is a challenge walking trail to the peak of Bartle Frere.

Babinda Boulders and Josephine Falls

The Wooroonooran National park also includes Josephine Falls at it's base, which includes a lovely series of cascades and a natural rockslide. Just north is Babinda Boulders, a picnic and swimming spot.

The Atherton Tablelands

The Atherton Tablelands are located about an hour and a half drive from Cairns. There are several roads entering the region, but they all require travelling up and over the great escarpment. Due to the higher altitude, a typical day on the tablelands is cooler and less humid than in Cairns and on the coastal plain. These tablelands were the site of some of Australia's most recent volcanic activity and past lava flows have resulted in the pleasant undulating hillsides that are known as the tablelands. This has also resulted in rich basaltic soils, and so much of the area has been developed for a huge range of agricultural activities. The diverse range of accommodation, farmland and forest provide rich and accessible wildlife viewing.

Millaa Millaa

The Atherton Tablelands are one of the best places in Australia for waterfalls. The combination of high rainfall and rugged topography results in many scenic drops. The small town of Millaa Millaa is located near the 'waterfall circuit' a sealed road that visits three magnificent waterfalls within short distance of each other. Milla Millaa falls are probably the prettiest falls, and one of the most photographed images of the Wet Tropics (right).

Giant trees

There are several impressive specimens of large rainforest trees on the tablelands. At Lake Barrine are the 'twin' Kauri pines. On the Danbulla Drive that circles Lake Tinaroo is the 'Cathedral' Fig tree, and just outside the village of Yungaburra is the 'Curtain' fig.

Crater Lakes National Park

Volcanic activity resulted in the formations of the crater lakes Eacham and Barrine. They were formed at similar times in close proximity and today constitute the 'Crater Lakes National Park'. Lake Eacham is slightly smaller than Lake Barrine and less developed. There is a walk that goes around the entire lake, picnic areas and national parks offices nearby. The edges of the lake are good for turtles and water dragons. Lake Barrine has also a walking track around the perimeter, boat cruises, gardens, souvenir shop and a cafe with great views of the lake.

Cairns

Cairns is located in the geographic centre of the north Queensland wet tropical rainforest region, therefore it is the perfect jumping off point for the surrounding rainforests south, north and west. However there is also some great tropical rainforest in closer proximity. Within the town itself, there is forest around the Botanical Gardens as well as several trails that climb up Mt.Whitfield. Behind Cairns there is a short drive to Crystal Cascades, and slightly further out behind the city is the Copperlode Dam area.

Kuranda and Barron Gorge National Park

Just a ten minute drive north of Cairns is the Barron River. This river begins at high altitude, and then winds through rainforest and agricultural land of the Atherton Tablelands. It passes by Kuranda and was responsible for cutting through the escarpment country to create the Barron Gorge. At the top of the gorge are Barron Falls. These falls have been an attraction for over a century, and were one of the first tourist stops in the region. The once mighty flow is today diverted for hydroelectricity, so now the full falls are only seen in flow during the extreme wet season, (pictured below). In such a flow, the water tumbles over 260 meters to the river below. The waterfalls, and much of that river, are protected as part of the 'Barron Falls National Park'. The best known of the views of the Barron Falls is from above the Barron Falls railway station on the Cairns Kuranda railway. There are also views of the gorge from the Skyrail.

The Daintree Region

The name 'Daintree' is used to refer to several different areas, including the Daintree River, a small town, a national park, and the general region.

The latter is often used to describe much of the area north from the Daintree River to just south of the Bloomfield Valley, however the actual 'Daintree National Park' is slightly different to this general perception and covers a large area south of the Daintree River. It is divided into several sections, the two largest being the 'Mossman Gorge section' and the 'Cape Tribulation section', with the Daintree River roughly in between the two.

The Mossman Gorge section is actually the largest part of the Daintree National Park, at over 55,000 hectares. It includes the Mossman River, the gorge up the river, and some of the mountains, such as Mt.Demi. Right at the entrance of the park is the Mossman Gorge Aboriginal Community, where guided walks on Aboriginal life in the rainforest are conducted. The park provides a nice, easy and accessible introduction to tropical rainforest. Ideally, it is best experienced first thing in the morning, before too many people arrive. Avoid hot afternoons, especially on weekends.

The Daintree River itself wasn't named after a type of tree, but a man named Richard Daintree. He was a geologist who worked as a government surveyor, although he reportedly didn't explore the area. The ferry operates from 6am to midnight and there is a fee to cross. There are many tours that explore the river, leaving from the river crossing and the Daintree village.

The 'Cape Tribulation section' is on the northern side of the Daintree River. At over 16,000 hectares in area it is actually smaller than the Mossman Gorge section. It includes Cape Tribulation and some of the coast, some of the Cape Tribulation road, many of the scenic mountains to the west of the road, such as Thornton Peak, and much of the forest along the coast towards the Bloomfield. The road up to Cape Tribulation itself is now sealed and does not require a 4WD except during times of high rain and flooding. However, a visit does require time. As the Daintree is a large area, and Cape Tribulation is at least 3 hours drive from Cairns, to make the most of a trip it is recommended that you stay for a few days. If your time is limited, then an organised day tour from Cairns or Port Douglas is probably the best option. There is a range of accommodation in the region, including backpacker hostels and more upmarket lodges. There are only limited chances for hiking or bush-walking in the Daintree, however, there is a good selection of well-made and maintained boardwalks.

Cape York

North of the 'wet tropics' of North-east Queensland begins Cape York. Generally this region is dominated by *Eucalyptus* woodlands, *Melaleuca* wetlands, and mangroves (all these are subjects of forthcoming Ecosystem Guides). However, there are patches of tropical and monsoon rainforest which are of interest as they hold some different species, and are also closer again in affinity with the great jungles of New Guinea. The best place for diversity and quality rainforest is no doubt Iron Range National Park, however, there are many other smaller patches of rainforest, monsoon forest and riparian forest applicable to this guide. For example, at Cape Melville there exists monsoon forest in and around the bases of mountains of black granite boulders (pictured above), and even the most casual and superficial tourist to the top of Cape York needs to walk through rainforest on the boardwalk from the car-park to the tip. Generally, however, the region is difficult to access and travel around. Beyond Cooktown the roads are 4WD, bumpy and hard going. There are many companies that conduct tours up through the region in the dry season. Most take a week, but this short time means you will be travelling almost all day every day, so to see any of the rainforest such as the Iron Range requires shopping around to more specialised operators running tours of several weeks duration.

The Top End and the Kimberly

The 'top end' includes the top half of the Northern Territory. There are patches of monsoon forest such as around Berry Springs, and also throughout Kakadu and Arnhemland. Kakadu is accessible during the dry and all year around with a range of tours leaving from Darwin and Jabiru, a small town inside of Kakadu. In the north-west of Australia is the Kimberly region. Although it is dominated by quite harsh woodlands and grasslands, there are many small patches of monsoon forest, found mainly around the coast and escarpment areas. While the inland Kimberly is accessible in the dry and by tours all year round from Broome and Darwin, the coast is much more inaccessible and expensive to explore. There are an increasing number of cruise companies exploring the coast, such as the nature-based expedition tours by Coral Princess Cruises.

References

- Adam, P. (1997). *Australian Rainforests*, Oxford University Press, Oxford, New York, U.S.A.
- Allen, G.R. (1994) "Silversides and their allies" IN Paxton, J.R. and Eschmeyer, W.N. (eds). *Encyclopedia of fishes*, UNSW Press, Sydney, New South Wales, Australia.
- Allen, G.R., Midgley, S.H. and Allen, M. (2002). *Field Guide to the Freshwater Fishes of Australia*, CSIRO, Collingwood, Victoria, Australia.
- Anderson, A. N. (2000). The Ants of Northern Australia, CSIRO Publishing, Collingwood, Victoria, Australia.
- Archer, M., Hand, S.J. and Godthelp, H. (2000), *Australia's Lost World; Riversleigh, World Heritage Site*, Reed New Holland Publishers, Frenchs Forest, N.S.W., Australia.
- Augee, M. L. "Echidnas" IN McDonald, D. W. (editor) (2006). *The Encyclopedia of Mammals*, Oxford University Press, Oxford, United Kingdom.
- Banfield, E. J. and Noonan, M. (1994). *Confessions of a Beachcomber*, University of Queensland Press, St. Lucia, Queensland, Australia.
- Barker, J., Grigg, G.C. and Tyler, M.J. (1995). *A field guide to Australian Frogs*, Surrey Beatty and Sons, Chipping Norton, NSW, Australia.
- Beehler, B.M., Pratt, T.K. and Zimmerman, D. (1986). *Birds of New Guinea*, Princeton University Press, Princeton, New Jersey, United States of America.
- Behler, J.L. and Behler, D.A. (2005). *Frogs – a chorus of colours*, Sterling Publishing Co, New York, New York, United States of America.
- Bermingham, E., Dick, C. W., and Moritz, C. (editors) (2005). *Tropical Rainforests, past present and future*, The University of Chicago Press, Chicago, U.S.A.
- Bottoms, T. (2000). "Bama country – Aboriginal Homelands" IN (editors McDonald, G. and Lane, M.) *Securing the Wet Tropics?*, The Federation Press, Annandale, N.S.W., Australia.
- Bostock, P. and Field, A. (2007). "A living Museum" IN *Wet Tropics - world heritage area magazine*, Wet Tropics Management Authority, Cairns, Queensland, Australia.- Braby, M. F. (2004). *The Complete Field Guide to Butterflies of Australia*, CSIRO Publishing, Collingwood, Victoria, Australia.
- Breeden, S. (1999) *Australian World Heritage Tropical Rainforest*, Steve Parish Publishing, Archerfield BC, Qld, Australia.
- Burton, J. A. (1998) . *The Book of Snakes*, Australian Book Centre, Sumner Park, Queensland, Australia.
- Cannon, L. (2000) 'Worms' IN Ryan, M and Burwell, C. (editors),. *Wildlife of Tropical North Queensland*, Queensland Museum, South Brisbane, Qld, Australia.
- Cayley, N.W. and Lindsey, T.R. (2000), *What Bird is that?* Redwood Editions, Dingley, Victoria, Australia.
- Chin, W. Y. (1998). *Ferns of the tropics*, Kangaroo Press, East Roseville, N.S.W., Australia.
- Churchill, S. (1998). *Australian Bats*, Reed New Holland, Frenchs Forest, New South Wales, Australia.
- Clarke and Lee (1997), *Name that flower; the identification of flowering plants*, Melbourne University Press, Victoria, Australia.
- Clyne, D. (1977). *A Guide to Australian Spiders*, Thomas Nelson, West Melbourne, Victoria, Australia.
- Cogger, H.G and Zweifel, R.G. (1998) *Encyclopedia of Reptiles and Amphibians*, UNSW Press, N.S.W., Australia.
- Compton,S.G., Wiebes, J.T. and Berg, C.C. (1996). 'The Biology of fig trees and their associated animals', *Journal of Biogeography*, 23: 405-407.
- Cooper, W and Cooper, W. (1994) *Fruits of the Rainforest, a guide to fruits in Australian Tropical Rainforests*, Geo Productions, Chatswood, NSW, Australia.
- Cooper, W. and Cooper, W. (2004) *Fruits of the Australian Tropical Rainforest*, Nokomis Editions, Melbourne Victoria, Australia.
- Cranbrook, E. and Edwards, D. (1994). *A Tropical Rainforest: the nature of biodiversity in Borneo at Betalong, Brunei*, Sun Tree Publishing, Singapore and The Royal Geographic Society, London, United Kingdom.
- Cribb, A.B. and Cribb, J.W.(1985). *Plant Life of the Great Barrier Reef and adjacent shores*, University of Queensland Press, St. Lucia, Queensland, Australia.
- Crome, F.H.J. (1975). 'The Ecology of Fruit Pigeons in Tropical North Queensland', *Australian Wildlife Research*, 2: 155-85.
- Crome, F.H.J. 'Vertebrates and Successions' in Webb, L.J. and Kikkawa, J. *Australian Tropical Rainforests; science-values-meaning*, CSIRO Publications, East Melbourne, Victoria.
- Cronin, L. and Westmacott, M. (1989), Key Guide to Australian Palms, Ferns and Fern allies, Reed, Chatswood, N.S.W., Australia.
- Davie, P. JF. and Short, J. W. (2000). 'Crustaceans' IN Ryan, M. and Burwell, C. (editors), *Wildlife of Tropical North Queensland*, Queensland Museum, South Brisbane, Queensland, Australia.
- Doyle, T. (2000). "The campaign to save the wet tropics" IN (editors McDonald, G. and Lane, M.) *Securing the Wet Tropics?*, The Federation Press, Annandale, N.S.W., Australia.
- Driml, S. (2000) 'Ecotourism - opportunities and threats' IN McDonald, G. and Lane, M. (eds). *Securing the Wet Tropics?*, The Federation Press, Annandale, New South Wales, Australia.
- Edgar, G.J. (2000). *Australian Marine Life; the plants and animals of Temperate Waters*, Reed New Holland Publishers, Frenchs Forest, New South Wales, Australia,
- EOTAW (1977), *Encyclopedia of the Animal World*, Bay Books, Sydney, N.S.W., Australia.
- Ermert, S. and Clapp, L. (1998). *Gardener's Companion to Weeds*, Landsdowne, Sydney, N.S.W., Australia.
- Farrant, P.A. (1999). Colour in Nature, Blandford, London, U.K.- Ferrett, R. (2005). Australia's Volcanoes, Reed New Holland, Frenchs Forest, Sydney, N.S.W., Australia
- Figgis, P. (1992). *Rainforests of Australia*, Ure Smith Press/Weldon International, Sydney, N.S.W., Australia.
- Flannery, T (1995a). *Mammals of New Guinea*, Reed Books, Chatswood, N.S.W., Australia.
- Flannery, T (1995b). *Mammals of the South-west Pacific and Moluccan Islands*, Reed, Chatswood, N.S.W., Australia.
- Fry, C. H. (1996). "Kingfishers" in Perrins, C. (ed) *The New Encyclopedia of Birds*, Oxford University Press, Oxford, United Kingdom.
- Fuentes. M. (2000). 'Frugivory, seed dispersal and plant community ecology', *Trends in Ecology and Evolution*, 15 (12): 487-488.
- Fuhrer, B. (2001). *Field Companion to Australian Fungi*, Bloomings Books, Hawthorn, Victoria, Australia.
- Fuhrer, B. (2005). A *Field Guide to Australian Fungi*, Bloomings Books, Melbourne, Victoria, Australia.

- Gooderham, J. and Tsyrlin, E. (2002). *The Waterbug Book*, CSIRO Publishing, Collingwood, Victoria, Australia.
- Goosem, S. (2000). "Landscape, Ecology and Values" IN (editors McDonald, G. and Lane, M.) *Securing the Wet Tropics?*, The Federation Press, Annandale, N.S.W., Australia.
- Green, R.J. (1993). 'Avian Seed Dispersal in and near Subtropical Rainforests', *Australian Wildlife Research*, 20: 535-57.
- Grice, A. C. (1996). 'Seed Production, dispersal and germination in Cryptostedia grandiflora and Ziziphus mauritiana, two invasive shrubs in tropical woodlands of northern Australia', *Australian Journal of Ecology*, 21: 324-331.
- Grieg, D. (1999). *Field Guide to Australian Wildflowers*, New Holland Publishers, Frenchs Forest, N.S.W., Australia.
- Greer, A.E. (1997). The Biology and Evolution of Australian Snakes. Surrey Beatty and Sons, Chipping Norton, NSW.
- Gullan, P.J. and Cranston, P.S. (1996), *The Insects; an Outline of Entomology*, Chapman and Hall, London, U.K.

- Hacker, J.B. (1990). 'Drift seeds and fruit on Raine Island, northern Great Barrier Reef, Australia', *Journal of Biogeography*, 17: 19-24.
- Hall, L. and Richards, G. (2000). *Flying Foxes, fruit and Blossom Bats of Australia,* University of New South Wales, Sydney, N.S.W., Australia.
- Hangay, G. and German, P. (2000). *Insects of Australia, Nature Guides,* New Holland Publishers, Frenchs Forst, N.S.W., Australia.
- Harborne, J.B. (1997), *Introduction to Ecological Biochemistry*, Academic Press, London, U.K.
- Harvey, M. S. and Yen, A. L. (1995). *Worms to Wasps, an illustrated guide to Australia's terrestrial invertebrates,* Oxford University Press, South Melbourne, Victoria, Australia.
- Heywood, V.H. (1978). *Flowering Plants of the World,* Oxford University Press, Oxford, United Kingdom.
- Heywood, V.H., Brummitt, R.K., Culham, A., and Suberg, O. (2007). *Flowering Plant families of the World,* Kew Botanical Gardens, United kingdom.
- Herbert, B. and Peeters, J. (1995), *Freshwater Fishes of Far North Queensland*, Department of Primary Industries, Brisbane, Queensland, Australia. Horne, P. and Crawford, D. (2005).
- *Backyard Insects*, The Miegunyah Press, Melbourne University Publishing, Carlton, Victoria, Australia.
- Hutton, W. and Cassio, A. *Tropical Fruits of Malaysia and Singapore*, Periplus Editions, Singapore.
- Hyland, B.P.M., Whiffin, T, Chistophel, D.C., Gray, B., Elick, R.W. and Ford, A.J. (1999). *Australian Tropical Rain Forest Trees and Shrubs*, computer key, CSIRO Publishing, Collingwood, Victoria, Australia.

- Ingrouille, M.J. and Eddie, B. (2006). *Plants - diversity and evolution*, Cambridge University Pess, Cambridge, United Kingdom.

- Jackes, B. and Cairns, A. (2001). *Plants of the Tropics*, James Cook University, Qld, Australia.
- Jarman, S. J. and Fuhrer, B. A. (1995). *Mosses and liverworts of rainforest in Tasmania and south-eastern Australia*, CSIRO Publishing, East Melbourne, Victoria, Australia.
- Jones, D. (1986). *Rainforest ornamental Plants in Australia*, Reed Books, Frenchs Forest, N.S.W., Australia.
- Jones, D. (1995), *Palms of the world,* Reed New Holland Books, Chatswood, N.S.W., Australia.
- Jones, D. (1998), *Cycads of the world,* Reed New Holland Books, Frenchs Forest, N.S.W., Australia.
- Jones, D. and Clemesha, S.C. (1981), *Australian Ferns and fern-allies,* Reed Books, Terrey Hills, Sydney, N.S.W., Australia.
- Jones, R.E. and Crome, F.H.J. (1990). 'The Biological Web- Plant/Animal Interactions in the Rainforest' in Webb, L.J. and Kikkawa, J. *Australian Tropical Rainforests; science-values-meaning*, CSIRO Publications, East Melbourne, Victoria.
- Jones, G. (2006). "Bats" IN McDonald, D. W. (editor) *The Encyclopedia of Mammals*, Oxford University Press, Oxford, United Kingdom.
- Juniper, T. and Serpell, J.A. (2003). 'Parrots, Lories and Cockatoos', in Perrins, C. (ed) (2003) T*he New Encyclopedia of Birds*, Oxford University Press, Oxford, United Kingdom.

- Kalko, E.K.V., Herre, E.A. and Handley, C.O. (1996). 'Relation of fig characteristics to fruit-eating bats in the New and Old World tropics', *Jounal of Biogeography*, 23: 565-576.
- Kantvilas, G. and Jarman, S. J. (1999). *Lichens of rainforest in Tasmania and southeastern Australia*, The Australian Biological Resources Study, Canberra, Australian Capital Territory, Australia.
- Kemp, A. and Evans S.M. (2003) IN Perrins, C. (ed). *The New Encyclopedia of Birds*, Oxford University Press, Oxford, United Kingdom.
- Keto, A. (1992). "Tropical Rainforests" IN Figgis, P. (editor). *Rainforests of Australia*, Ure Smith Press/Weldon International, Sydney, N.S.W., Australia.
- Kroon, F. and Westcott, D. (2001). 'Helmet Heads', *Australia Nature*, 26 (12): 62-69.

- Lack P. and Gill, F. B. (2003) IN Perrins, C. (ed). *The New Encyclopedia of Birds*, Oxford University Press, Oxford, United Kingdom.
- Larsen, K., Ibrahim, H., Khaw, S.H., and Saw, L. G. (1999). *Gingers of Peninsular Malaysia and Singapore*, Natral History Publications (Borneo), Kota Kinabalu.
- Law, B.S. and Spencer, H. J. (1998) 'Common Blossom Bat' IN Strahan, R. (ed), *The Mammals of Australia*, Reed/ New Holland Publishers, Frenchs Forest, N.S.W., Australia
- Lindsey, T. (1998). 'Starlings and their allies' in Forshaw, J. (ed) *The Encyclopedia of Birds,* UNSW Press, University of New South Wales, Sydney, N.S.W., Australia.
- Lindsey, T. (1998), *Green Guide to Parrots of Australia*, New Holland Publishers, Frenchs Forest, N.S.W., Australia.
- Loiselle, B.A. and Blake, J.G. (1999). 'Dispersal of Melastome Seeds by fruit-eating birds of Tropical Forest Understorey', *Ecology,* 80 (1): 330-336
- Lott, R.H., Harrington, G.N., Irvine, A.K. and McIntyre, S. (1995). 'Density-dependent Seed Predation and Plant Dispersion of the Tropical Palm *Normanbya normanbyi*', *Biotropica*, 27 (1): 87-95.
- Lovelock, C. (1993). *Field Guide to the Mangroves of Queensland*, Australian Institute of Marine Science, Townsville, Queensland, Australia.
- Low, T. (1991). *Wild Food Plants of Australia*, Angus and Robertson/HarperCollins Publishers, Pymble, Sydney, NSW, Australia.

- Martin, S. (1992-). *Tropical Topics; a compilation*, Wet Tropics Management Authority, Qld, Australia.
- Martin, S. (1993), 'Wet Tropics Webs 2', *Tropical Topics*, Queensland Department of Environment and Heritage, Cairns, Queensland, Australia.
- Martin, S. (1995) Tropical Topics; Where land meets sea- islands, cays and seabirds, *Tropical Topics,* Queensland Department of Environment and Heritage, Cairns, Queensland, Australia.
- Martin, S. (2000) *Tropical Topics; a second compilation*, Wet Tropics Management Authority, Queensland, Australia.
- Martin, S. (1997), 'Macropods of the Wet Tropics', 'Wet Tropics Geology', *Tropical Topics*, Queensland Department of Environment, Cairns, Queensland, Australia.
- Martin, S. (2002), 'Invisible recyclers', *Tropical Topics*, Queensland Department of Environment, Cairns, Queensland, Australia.
- Mattison, C. (1995). *Encyclopedia of Snakes*, Readers Digest, Surry Hills, N.S.W., Australia.

- McDonald, D. W. (editor) (2006). *The Encyclopedia of Mammals*, Oxford University Press, Oxford, United Kingdom.
- McDonald, G. and Lane, M. (2000). *Securing the Wet Tropics?*, The Federation Press, Annandale, New South Wales, Australia.
- McJannet, D. and Reddell, P. 'Where earth meets sky: cloud forests of the Wet Tropics', *Using Rainforest Research*, CRC for Tropical Rainforests Ecology and Management.
- Mead, C. J. and Lundberg, A. (2003) IN Perrins, C. (ed). *The New Encyclopedia of Birds*, Oxford University Press, Oxford, United Kingdom.
- Menkhorst, P. and Knight, F. (2001). *A Field Guide to Mammals of Australia*, Oxford University Press, South Melbourne, Victoria, Australia.
- Mirtschin, P. and Davis, R. (2000), *Snakes of Australia*, Hill of content, Melbourne, Victoria, Australia.
- Morris, I. (1999) *Steve Parish Natural History Guide: Kakadu National Park*, Australia, Steve Parish Publishing, Archerfield, Queensland, Australia.
- Moritz, C. (2005). "Overview: the Australian wet tropics" IN Bermingham, E., Dick, C. W., and Moritz, C. (editors) *Tropical Rainforests, past present and future*, The University of Chicago Press, Chicago, U.S.A.

- Nadolny, C. (1999). 'Recruitment and survival of *Argyrodendron actinophyllum* seedlings in an Australian rainforest', *Australian Journal of Ecology*, 24: 258-269.
- Nicholson, N. and Nicholson, H. (1988-2004). *Rainforest Plants I-VI*, Terania Publishing, The Channon, NSW, Australia.
- Nielson, L. (1997). *Daintree, Jewel of tropical North Queensland*, Published by Lloyd Nielsen, Mt.Molloy, Queensland, Australia.
- Norris, K. (1991), 'General Biology', CSIRO, *Insects of Australia*, Melbourne University Press, Melbourne, Victoria, Australia.
- Nott, J. (2005). "The origin and evolution of Australia's eastern highland" IN Bermingham, E., Dick, C. W., and Moritz, C. (editors) *Tropical Rainforests, past present and future*, The University of Chicago Press, Chicago, U.S.A.

- Obst, F. J. (1998) 'Turtles & Tortoises' IN Cogger, H.G and Zweifel, R.G. (editors) *Encyclopedia of Reptiles and Amphibians*, UNSW Press, N.S.W., Australia.
- Olsen, P. (1998) 'Owls, Frogmouths & Nightjars' in Forshaw, J. *The Encyclopedia of Birds*, UNSW Press, University of New South Wales, Sydney, N.S.W., Australia.
- Olsen, P. (1998). "Water Rat' IN Strahan, R. (ed), *The Mammals of Australia*, Reed/ New Holland Publishers, Frenchs Forest, N.S.W., Australia.
- Osborne, P.L. (2000). *Tropical Ecosystems and Ecological Concepts*, Cambridge University Press, Cambridge, United Kingdom.
- Oxford (2004). *The Australian Oxford Dictionary*, Oxford University Press, South Melbourne, Victoria, Australia.

- Parsons, M. (1999). *The Butterflies of Papua New Guinea: their systematics and biology*, Academic Press, London, U.K.
- Paxton, J.R. and Eschmeyer, W.N. (1994). *Encyclopedia of fishes*, UNSW Press, Sydney, New South Wales, Australia.
- Perrins, C. (ed) (2003). *The New Encyclopedia of Birds*, Oxford University Press, Oxford, United Kingdom.
Pestana, B. (1993). *Australian Agriculture; the complete reference on rural industry*, National Farmers Federation, Morescope Publishing, Camberwell, Victoria, Australia
- Pettigrew, J.D. (1987). 'Are Flying Foxes (Chiroptera: Pteropodidae) really primates?', *Australian Mammalogy*, 10: 119-124.
- Pike, G. (1987). *The Wilderness Coast*, Pinevale Publications, Mareeba, Queensland, Australia.
- Pizzey, G. (1985). *A Separate Creation*, Currey O'Neil Ross, South Yarra, Victoria, Australia.
- Pizzey, G. and Knight, F. (2003, 7th edition). *The Field Guide to the Birds of Australia*, HarperCollins Publishers, Pymble, Sydney, N.S.W, Australia.
- Preston-Mafham, R and Preston-Mafham K. (1998), Spiders of the World, Blanford, London, U.K.
- Primack, R. and Corlett, R. (2005). *Tropical Rain Forests: an ecological and biogeographical comparison*, Blackwell Publishing, Malden, MA, U.S.A.
- Purvis, W. (2000). *Lichens*, The Natural History Museum, London, United Kingdom.
- Pusey, B., Kennard, M. and Arthington, A. (2004). *Freshwater Fishes of North-eastern Australia*, CSIRO Publishing, Collingwood, Victoria, Australia.

- Raven, P.H., Evert, R.F. and Eichorn, S.E. (1992). *Biology of Plants*, Worth Publishers, New York, New York, U.S. A.
- Raven, R. (2000). 'Spiders' IN Ryan, M. and Burwell, C. (editors), *Wildlife of Tropical North Queensland*, Queensland Museum, South Brisbane, Queensland, Australia.
- Rentz, D. (1996). *Grasshopper Country*, UNSW Press, Sydney, N.S.W., Australia.
- Richards, G.C. (1987). 'Aspects of the ecology of Spectacled Flying Fox *Pteropus conspicillatus* (Chiroptera: Pteropodidae) in Tropical Queensland', *Australian Mammalogy*, 10: 87-88.
- Richards, P.W. (1998). *The Tropical Rain Forest (an ecological study)*, Cambridge University Press, Cambridge, Great Britain.
- Ritchie, R., Curtis, P., Keto, A., Scott, K. and Werren, G. (1995). *North Queensland Wet Tropics, a guide for travellers*, Rainforest Publishing, Millers Point, N.S.W., Australia
- Roberts, J., Fisher, C. and Gibson, R. (1995). *A Guide to Traditional Aboriginal Rainforest Plant Use*, Bamanga Bubu Ngadimunku, Mossma, Queensland, Australia.
- Robinson, R. (2003) in Perrins, C. (ed) *The New Encyclopedia of Birds*, Oxford University Press, Oxford, United Kingdom.
- Rohwer, J. G. (2002). *Tropical Plants of the World*, Sterling Publishing, New York, New York, United States of America.
- Rowell, R.J. (1996). *Ornamental Conifers for Australian Gardens*, University of New South Wales Press, Sydney, N.S.W., Australia.
- Ruppert, E.E. and Barnes, R.D. (1994). *Invertebrate Zoology*, Saunders College Publishing, Orlando, Florida, United States of America.
- Ryan, M and Burwell, C. (editors), (2000). *Wildlife of Tropical North Queensland*, Queensland Museum, South Brisbane, Queensland, Australia.

- Scarth-Johnson, V. (2000), *National Treasures; flowering plants of Cooktown ad Northern Australia*, Vera Scarth-Johnson Gallery Association, Cooktown, Queensland, Australia.
- Schoonhoven, L.M., Jermy, T. and van Loon, J.J.A. (1998), *Insect-Plant Biology*, Chapman and Hall, London, U.K.
- Shine, R. (1998). "Snakes" IN Cogger, H.G and Zweifel, R.G. (editors) *Encyclopedia of Reptiles and Amphibians*, UNSW Press, N.S.W., Australia.
- Shine, R. (1999). *Australian Snakes – A Natural History*, Reed New Holland Publishers, Frenchs Forest, N.S.W., Australia.
- Silsby, J. (2001). *Dragonflies of the World*, CSIRO Publishing, Collingwood, Victoria, Australia.
- Slater, P., Slater.P. and Slater, R. (2001) *The Slater Field Guide to Australian Birds*, Landsdowne Publishing, The Rocks, NSW, Australia.
- Smith, J. (1999) *Australian Driftseeds; a compendium of seeds and fruits commonly found on Australian Beaches*, University of New England, Armidale, N.S.W., Australia.

- Snow, D.W. (1981). 'Tropical Frugivorous Birds and Their Food Plants: A World Survey', *Biotropica*, 13: 1-14.
- Stocker, G.C. and Irvine, A.K. (1983). 'Seed Dispersal by Cassowaries (*Casuarius casuarius*) in North Queensland's Rainforests', *Biotropica*, 15: 170-176.
- Smith, J.M.B., Heatwole, H. ,Jones, M. and Waterhouse, B.M. (1990). 'Drift disseminules on cays of the Swain Reefs, Great Barrier Reef, Australia', *Journal of Biogeography*, 17: 5-17.
- Stanisic, J. (2000). "land Snails' IN Ryan, M and Burwell, C. (editors),. *Wildlife of Tropical North Queensland*, Queensland Museum, South Brisbane, Queensland, Australia.
- Stewart, D. (1998). *Australian Frog Calls: Tropical North-east*, Nature Sound, Mullumbimby, NSW, Australia.
- Strahan, R. (1998), *The Mammals of Australia*, Reed/ New Holland Publishers, Frenchs Forest, N.S.W., Australia.

- Terrel, E. and Ledger, J. (2000), *Daintree River to Cae Tribulation; a handbook for Tour Guides*, Daintree Rescue Program, Wet Tropics Management Authority, Cairns, Queensland, Australia.
- Theischinger, G. and Hawking, J. (2006). The Complete Field Guide to Dragonflies of Australia, CSIRO Publishing, Collingwood, Victoria, Australia.
- Tidemann, C.R.and Nelson, J.E. (1987). 'Flying Foxes (Chiroptera: Pteropodidae) and Bananas: some interactions', *Aust. Mammal.* 10: 133-135.
- Tomlinson, P.B. (1994). *The Botany of Mangroves*, Cambridge University Press, New York, New York, U.S.A.
- Torr, G. (1997). 'Forest Dragons', *Australia Nature*, 25(8): 32-39.
- Tracey, J.G. (1987), *The Vegetation of the Humid Tropical Region of North Queensland*, CSIRO, Melbourne, Victoria, Australia.
- Trott, L. (1996). *Wet Tropics in Profile*, the Wet Tropics Management Authority, Cairns, Qld, Australia.
- Turner, I.M. (2001). *The ecology of trees in the tropical rainforest*, Cambridge University Press, Cambridge, United Kingdom.

- Uchida, D. (1996) *Malaysia's Emerald Crown, exploring the world's oldest tropical rain forest,* Tropical Press, Kuala Lumpur, Malaysia.

- Valentine, P., Frith, C. and Frith, D. (1991), *Australian Tropical Butterflies*, Frith and Frith Books, Malanda, Qld, Australia.

- Warren, W., Young, J. and Barnard, L. (1999). 500 *Tropical Plants, for Australian Gardeners*, Random House, Milsons Point, N.S.W., Australia.
- Wehling, W.F. and Thompson, J.N. (1997), Evolutionary conservatism of oviposition preference in a widespread polyphagus insect herbivore, Papilio zelicaon, 111; 209-215.
- White, M. (1998). *The Greening of Gondwana*, Simon and Schuster, East Roseville, N.S.W., Australia.

- Whitmore, T.C. (1999). *An Introduction to Tropical Rain Forests*, Oxford University Press, Oxford, New York, U.S.A.
- Whitten, A.J., Damanik, S.J., Anwar, J. and Hisyam, N. (1987). *The Ecology of Sumatra*, Gadjah Mada University Press, Yogyakarta, Indonesia.
- Whittaker, R.J. and Jones, S.H. (1994). 'The role of frugivorous bats and birds in the rebuilding of a tropical forest ecosystem, Krakatau, Indonesia', *Journal of Biogeography*, 21, 245-258.
- Whybird, O. and Clague, C. (2001). 'November 2000 Spectacled Flying-fox Survey', *Flying Fox Count Report*, Phoniscus, Millaa Millaa, Queensland, Australia.
- Wilson, B. (2002). *A Handbook to Australian Seashells (on seashores East to West and North to South)*, New Holland Publishers, Frenchs Forest, New South Wales, Australia.
- Wilson, M.F., Irvine, A.K. and Walsh, N.G. (1989). 'Vertebrate Dispersal Syndromes in Some Australian and New Zealand Plant Communities, with Geographic Comparisons', *Biotropica*, 21 (2): 133-147.
- Wilson, S. and Swan, G. (2003). *A Complete Guide to the Reptiles of Australia,* Reed New Holland, Frenchs Forest, New South Wales, Australia.

- Young, T. (2005). *A field guide to the fungi of Australia*, UNSW Press, Sydney, New South Wales, Australia.

- Zborowski, P. and Edwards, R. (2007) *A Guide to Australian Moths*, CSIRO, Collingwood, Victoria, Australia.
- Zborowski, P. and Storey, R. (1995) A Field Guide to Insects of Australia, Reed New Holland, Frenchs Forest, N.S.W., Australia.

Index